GAYLORD MG

Rapture

RAPTURE

How Biotech Became the New Religion

BRIAN ALEXANDER

Basic Books
A Member of the Perseus Books Group
New York

Copyright © 2003 by Brian Alexander
Published by Basic Books,
A Member of the Perseus Books Group

Library of Congress Cataloging-in-Publication Data

Alexander, Brian, 1959–
  Rapture : how biotech became the new religion / Brian Alexander.—
1st ed.
    p. cm.
  ISBN 0–7382–0761–6 (alk. paper)
  1. Biotechnology—Popular works.   2. Molecular biology—Popular
works.   I. Title.

  TP248.215.A43 2003
  303.48'3—dc21   2003010561

Text design by Bookcomp, Inc.
Set in 10.5-point Fairfield by Perseus Publishing Services

First Edition
1 2 3 4 5 6 7 8 9 10—06 05 04 03

for Shelley

Contents

1   Waiting for the Rapture   1

2   The Prophet   11

3   The Endless Frontier   27

4   Arise, Lazarus Long!   47

5   The Immortal Mr. Steinberg   65

6   Way Out West   103

7   Bring On the Inquisition   125

8   Water into Wine   159

9   Pop! Goes the Rapture   201

10   The Rapture Rides in a Limo   223

Acknowledgments   259

Notes   261

Index   279

The biologist is the most romantic figure on earth at the present day.

—J.B.S. HALDANE, 1923

I sometimes regret not to believe in God. It would be so comfortable to rely upon him, at least as long as medicine doesn't progress a little further.

—SALVADOR LURIA, 1944

In my opinion in the year 2001 so many physical problems will have been surmounted that a woman's beauty will be a dream that will be completely obtainable. Obviously ravaging disease will have been surmounted.

Obviously the problem of shape and line will be so controlled that a woman will have a strong supple and splendid body all her life . . .

It will then amuse and entertain her to paint herself as if she was a heathen idol . . .

She will walk like a dancer . . .

Her eyelashes will be shiny and delicious the way Garbo's were long before there were false eyelashes . . .

Any woman who is not looking forward to the future isn't thinking clearly because today we are between two worlds craving leisure because we feel it will give us more time for beauty and craving to be busy and productive. Then with a splendid circulation and clear eye and mind that is working easily and doesn't have pressure on it will all be alleviated. The future holds a golden world.

—DIANA VREELAND, 1967

I   WAITING FOR THE RAPTURE

DECEMBER 1994. Alexis Park Hotel. Las Vegas, Nevada.

Ralph Merkle admitted that dunking your dead body into a tank filled with liquid nitrogen like a Krispy Kreme into a cup of Kona would have side effects. At about 196 Celsius degrees below zero, the stainless steel dewar would be pretty cold. So there could be damage. Ice crystals could form and puncture some of your body's cells. You could emerge from the deep freeze looking more like a long-frozen rib eye than a sleek cryonaut resurrected from a time warp stasis. On the other hand, what would you have to lose?

"This is the second worst thing that could happen to you," Merkle told the crowd. The audience, seated under a big white tent out by the pool, chuckled and nodded their heads. Yes, freezer burn beat the hell out of permanent death. Besides, this was Vegas, and if you wanted to win big, if you wanted to become immortal, you had to play long odds, right? So even though it sounded crazy, who knew? Maybe, in twenty or forty or seventy-five years, you'd be taken out of the vat. Maybe nanobots, teeny-tiny machines the size of a few atoms, would be sent in to cruise the byways of your body to repair the mushy bits. Then doctors would fix whatever it was that had killed you. Bad heart? You'd get a new one, then a jump-start, a pat on the behind, and out the door you'd go to revel in your resurrection.

The Second Annual Conference on Anti-Aging Medicine was all about big gambles and big payoffs. The confab was produced by a group called the American Academy of Anti-Aging Medicine (A4M), founded by a couple of Chicago-area osteopaths, Ronald Klatz and Bob Goldman. Both had long been active in the shadowy realms of life extension and body building. Klatz, for instance, served as an on-site medical doctor for the El Dorado Rejuvenation and Longevity Institute, a Cancún clinic that dispensed human growth hormone (HGH) injections. Now Klatz and Goldman were taking their life extension message to the broader world. Whereas most of the Cancún clients were wealthy, among the few who could afford to fly to Mexico and pay for the expensive shots, by forming A4M and coming to Vegas, Klatz and Goldman were now attracting a large crowd of folk, some rich, many not. They were lawyers and engineers and unreconstructed hippies. They came from cities and towns across the country. But all of them, 200 or so, were united in one belief: death was just damn unfair.

Merkle, then a scientist at Xerox's renowned Palo Alto Research Center, and a founding father of theories behind nanotechnology, the science of the extremely small, proved his own belief by wearing a bracelet advising medical workers that in the event of his impending death they should contact the Alcor Life Extension Foundation, the world's leading cryonics outfit. Merkle, an adviser to Alcor, admitted that cryonic suspension sounded far-fetched, and he appeared to look upon the prospect as an experiment. But the people under the tent were willing to take the ride. If all their other strategies—the fitness routines, the vitamins and herbs, the designer chemicals—failed to provide the superlong life they hoped to achieve, cryonics seemed like a perfectly reasonable bailout option. True, at around $120,000, freezing your whole body would be expensive, but there was a head-only option for $50,000. For that price, Alcor technicians would cut your head off, cap your neck, fill your skull with antifreeze, and then submerge it in a stainless condominium with other "neuros."

If all went according to plan, you would be thawed and cured and, if necessary, a new body would be cloned for you. True, no mammal had ever been cloned. But the cryonics folks were pretty sure it would happen someday. They had been weaned on science fiction and raised with an unalloyed belief in the future. They firmly believed that science was an engine of miracles, that given enough time, science could make the wildest dreams come true.

Alas, their love was unrequited. Mainstream science, especially the biological sciences, shunned them, embarrassed by the ways in which the life extensionists and those who eagerly anticipated an augmented, enhanced human future hungrily grabbed hold of the slightest of developments and used them to trumpet an impending biorapture. The career currency of mainstream science is reputation, and associating with people who talked about nutty ideas like extending the human life span or giving human beings a new set of enhanced abilities, was a good way to go bankrupt. "You have to approach the outside world very cautiously when you present some of these ideas," Stephen Coles, M.D., Ph.D., told the crowd during his talk on co-enzyme Q10, a supplement he reckoned was an elixir. "In this audience I hope I can come out of the closet."

Life extensionists imagined themselves an enlightened, if perse-cuted, minority. "Big medicine" was out to stifle their innovative ideas. Drug company conspiracies prevented knowledge of lifesaving dietary supplements from reaching consumers. The Food and Drug Administration (FDA) wielded unreasonable demands for experi-mental data like weapons to cut down the age-defying properties of substances the life extension community had uncovered through its own hard work, substances most of them were using right now. The Life Extension Foundation (LEF), an advocacy group started by an antiaging zealot named Saul Kent, even set up a display advertising a new attraction meant to draw visitors to its Florida headquarters: the FDA Holocaust Museum. The FDA was killing thousands every year, the foundation insisted, by standing in the way of the therapies championed by the LEF. The government, the life extensionists believed, was part of the culture of death. It *wanted* us to die!

Such whining was part of the life extension territory, born from decades of frustration. Why couldn't the rest of the world see? Death could be broken. Human beings could be made better than they are now—smarter, stronger, healthier. But here they were, out in Las Vegas, fighting the lonely fight while the rest of the world sheepishly accepted death and human limitation. And what did they get for their trouble? Ridicule. Government harassment.

Still, they were undeterred. Their evangelical mission was to convince as many people as possible, including mainstream science, that their cause was not only just, but possible. So they did what they could, promoting the few scientists who hinted that, perhaps, the life extensionists and the bioutopians were not crazy after all. At this meeting they were about to give the first-ever "Infinity Award" to Roy Walford, a UCLA scientist who used special low-calorie diets to extend the lives of lab mice, and who had gone on the diet himself. And Richard Cutler, a scientist from the National Institute on Aging, one of the National Institutes of Health (NIH), gave a life extension pep talk earlier in the day. He said that while "aging is ubiquitous," it could be attacked. "If we are sincere about increasing life span, we have to do something about the aging process itself, not just piece-meal diseases." Maybe, Cutler said, aging was in the genes. If so, then the question became "Can we intervene?" Everyone hung on to that question like a buoy in a dark sea because they all knew that a lot was happening with genes. The Human Genome Project had been going on for six years. Surely something would come of it.

This was the message they preached with all the energy their optimism could muster, making the deep dive into a tank of liquid nitrogen sound like the most practical thing in the world. After all, despite the uncertainties of cryonics, science—especially biology and genetics—was on the threshold of great discoveries. Those discoveries were coming soon and if you were dead and indisposed, your veins filled with embalming fluid or your bones turned to ashes by the cremator's flames, that would be a tragedy, the ultimate missed opportunity. So it was best to prepare. At least one woman in the audience was convinced.

"I think I am going to sign up to have my head frozen," she told her female companion.

But the other foresaw trouble. "Just the head? I dunno."

"This old body is going to be worn out by then anyway."

They both laughed.

"But what if they put you on a fat body?"

The first woman thought, then, with a can-do grin on her face, said, "No, that would never happen. I'd specify."

It was easy to sit in the crowd of life extension enthusiasts, to whom concepts like cryonics make for lunchtime conversation, and wonder why the strategies for achieving superlongevity had not moved much beyond cryogenic preservation, why cryonics was still a presence at a conference that was obviously aimed at achieving a level of acceptability for a goal—life span extension—that had always been considered science fringe. It was a seeming admission that hopes to modify human biology or to extend the human life span had not progressed beyond hunkering down in a deep freeze and waiting for better days. While the conference showcased many other age-defying strategies, from Eastern religious meditation and dietary plans, to nutritional supplement regimens calling for mega doses of vitamins and antioxidants and hormones like melatonin, and while Cutler sounded encouraging, extending the human life span was just as foggy a prospect as it had always been. Everyone tried to ignore the awful truth that there was no such thing as anti-aging medicine, that no drug, no diet, no vitamin, no exercise had ever been proven to make people live longer.

In their hearts, though, they had to know. All the effort, all the money, would not, in the end, save them. In America, if you work hard, obey the rules, and really, truly, believe, you are supposed to achieve your goals. Yet these people were going to die. They felt cheated. The desperation was written on the faces of many atten-dees, faces made gaunt and sallow by esoteric diets, faces that prompted a local PR woman hired by the organizers to whisper to me, "I wish all these people would just eat a fuckin' steak."

But these people were nothing if not stubborn. Despite the unstoppable ticking of the clock, everyone was determined to carry on. On the second afternoon of the conference, I walked outside the tent to the pool deck. There, Durk Pearson was standing like a may-pole amid a gaggle of life extensionists. In 1982, Pearson and wife Sandy Shaw had released their book, *Life Extension: A Practical Sci-entific Approach*. The words *practical* and *scientific* lent the title legitimacy and gave their system a profoundly democratic tone, reawakening in the minds of a very broad audience the idea that we could live longer, maybe far longer, than we had thought. The book was a hit because Pearson said that everybody, even you and me, could stave off the worst effects of aging by taking huge doses of cer-tain nutrients and chemicals and adhering to special dietary plans. Soon, Pearson and Shaw were extolling the benefits of vitamins and exotic-sounding nutrients on local newscasts around the country. They appeared on the Merv Griffin show. They made personal appearances. They created their own line of supplements marketed in health food stores and through catalogues, products to make you smarter, live longer, improve your sex life.

Pearson's star had faded just a bit since those heady days. Now he and Shaw lived in the Nevada desert, where they spent their time reading science journals, raising ball pythons, and haranguing the FDA for the jaundiced eye the agency cast on dietary supplements and the claims made for them. Still, among life extensionists, the tall, gangly Pearson remained an antiaging oracle, admired for the way he brought the life extension message into the living rooms of America.

So out on the pool deck, he was peppered with questions about the latest supplements.

"Do you take seratonin? How much? Where can I get it?"

Pearson began to feel crowded. He walked toward the hotel lobby door, only to be followed by fans shouting questions like White House reporters trying to be heard over chopper blades. Pearson dodged chaise lounges, zigzagging his way toward the door. Cut off, he doubled back again, all the while tossing nuggets.

"L-arginine . . . 300 milligrams . . . choline cocktail . . . free radicals . . ."

"The carotenes, Durk, the carotenes," one supplicant shouted urgently. "Do you take the carotenes?"

"Two hundred fifty thousand milligrams every day in my orange juice."

Carotenes give yellow squash and carrots their colors and this apparently concerned the man who had asked the question.

"Gee, Durk, isn't that a lot?"

Pearson halted, straightening his usually stooped frame. He stared directly into the man's eyes, waited a beat, and said, "I'd rather be orange than dead."

WILLIAM HASELTINE had never heard of Durk Pearson, though Pearson, a dedicated reader of science journals, had almost certainly heard of Haseltine. Even as Pearson was delivering his stark choice—orange or dead—Haseltine, a member of the biology elite, was at work, trying to deliver another option.

Two and half years before A4M's second annual meeting in Las Vegas, Haseltine had attended a confab of his own. This one was scheduled for the Metro stop in Bethesda, Maryland, and Haseltine was skeptical the rendezvous would amount to much. He had never met J. Craig Venter, but he knew a little about Venter's work and he knew others who had dealt with Venter over the years. The word was that Venter was unpleasant, abrasive—the same rap Haseltine lived with—and that Venter sometimes performed sloppy science. Besides, Venter had gone to graduate school at the University of California at San Diego, a new university in the years Venter was there, and far removed both physically and in terms of reputation from the select Harvard–MIT world Haseltine inhabited.

Haseltine was a star. His biology career had been mentored by one Nobel Prize winner after another, including James Watson, who, with Francis Crick, had been first to guess at the double helix structure of DNA. Haseltine had made quite a name for himself as one in a group of scientists who had unraveled some of the genetic mystery surrounding HIV, the virus that causes AIDs. He had his own lab at Harvard's Dana-Farber Cancer Center. He was pulling down government research grants. He got his name in the newspapers. He partied at AIDs fund-raisers in California, New York, Europe. He knew everybody who was anybody in the small world of biology in which he circulated and had gotten to know a lot of connected people in the worlds of politics and finance, including society types from up and down the East Coast. In recent years he had started making a lot of money from his foray into the world of corporate biotechnology and from advising a venture capital firm about its own biotechnology investments.

Venter was an ex-surfer who worked for a government salary at the NIH. He was a lab drone who had come up with a clever way, so he said, of finding genes within the chaos of the human genome. He wanted more money from the NIH to ramp up his gene-hunting lab, but he was asking for a lot of money, and besides, Venter had a reputation for raising hackles within the institutes, especially when it came to decisions about whether or not to patent his gene information. The NIH wasn't about to drop as much money as he required in his lap, so Venter had begun talking to industry. He had recently made contact with the venture capital firm that used Haseltine as an adviser. Venter and the firm were talking about a lot of money, $70 million, maybe $80 million, to take Venter away from the government and set him up in his own gene-hunting lab. So the investors asked Haseltine to fly down to Washington, take the Metro out to Bethesda to meet with Venter, and check out Venter's technology and his data. Haseltine protested mildly, but even though he thought the trip would be pointless, he agreed to go.

Venter and Haseltine made a date for lunch at the Metro stop. There was nothing special about the station. It was just a concrete slab in the center of town surrounded by the high-end shops that catered to the upwardly mobile. But you could get a burger and a soda at the cafeteria and, since it was a nice, warm day, Haseltine and Venter sat in the sun just outside the cafeteria for forty-five minutes, after which Haseltine picked up a few computer printouts Venter had given him, shook Venter's hand, climbed aboard the Metro train, and began to plan the ways in which his life would now be different.

Venter had handed Haseltine data showing that technicians using machines could produce a gusher of genetic information, much of it patentable. Along with the information came the proteins that genes instruct the body's cells to make. The venture firm thought it could form a company around this technology, tease out the genetic data, and then sell that data to big pharmaceutical companies desperate to lock onto new ideas for drugs. But why, Haseltine now thought, settle for small potatoes?

People who know Haseltine fall into two camps. One camp says he is a visionary. The other says he is a blowhard. Haseltine places himself firmly in the visionary camp. He has been a visionary of sorts since he was nine years old, in 1953, the very year, coincidentally, that Watson and Crick published their findings on the double helix. That was the year he decided he was going to change medicine someday. His mother had been ill, one thing after another, and though he wasn't sure how, exactly, he was certain he wanted to fix it so that people did not have to suffer the way his mother had. They shouldn't have to suffer at all and Haseltine determined that he would be one of the men to deliver human beings from suffering through the miracles of science. His whole life had been dedicated to this very proposition, and now he saw clearly how to take the next, biggest step. Sell data? Well, maybe for a while, to bring in some cash, but with the kind of results Venter was showing him, a company wouldn't have to sell information to big pharma so it could make proteins. A company could make proteins on its own, whenever it wanted and however much it wanted. Or, conversely, it could use the information to create an antibody to inhibit the effects of a protein, like a protein that cancer cells need for growth. It would not only have hundreds of possible drug candidates, but also thousands of the instructions cells send to each other and to themselves.

Why create just an information sales company? Why not also use the information to make your own drugs? "I saw a very clear opportunity to build a new company, to solve most of the key problems that most biotech companies had had. That is, too few ideas, too little money, and too little time. I felt, with this technology, we could get unlimited ideas for new drugs. One gene made one protein, made one drug or one antibody target. It was very simple."

But there was much more. Equipped with the instructions of life, Haseltine could finally begin to think about fulfilling his bigger ambition: to begin to eliminate human suffering, to take control of human biology and turn human beings from the evolved into the evolvers. Maybe he could even help defeat death.

This had been a grand old ambition of biology, though hardly anybody talked about it anymore. It sounded crazy, this business of making people live a long time, and the idea of altering human biology, of tinkering with it like a mechanical toy, conjured images of the beast men in Wells's *The Island of Dr. Moreau* in the minds of those who had come to see human genes as an inviolable legacy of nature or of God. But Haseltine was only thinking what some other scientists had thought years before and what some had begun to think again even as he met with Venter. He saw no reason in the world why he should not be one of those who stretched out his fingertip to touch the hand of man. It was as if fate had delivered him to that Metro stop, and before he was through, Haseltine would make himself a symbol of all that biotechnology could offer those who believed in the coming biorapture.

Some of them had been waiting years. They had, in their own ways, done the best they could. They read science journals and tried to make educated guesses. They consumed huge amounts of vitamins. They injected themselves with calf thymus extract and with cells from fetal sheep. They wrote to offshore pharmacies to obtain synthetic chemicals they hoped would enhance their brains. A few took wild flyers on research outside of academic dogma. Some took cocktails of hormones to enhance their bodies. Others placed hope in computers, thinking that they might someday save their very souls by storing them on disk drives.

But though the congregants in Las Vegas would not realize it in 1994, the gulf they felt so keenly, the vast distance between themselves and the mainstream of science and biotechnology, was already rapidly closing. In the coming years, Durk Pearson would take to quoting Haseltine and members of the life extension movement would call him a hero. Biotechnology would be seen not as an enemy, not as part of the conspiracy to stifle human possibility, but as the best chance man had ever had to escape the shackles of limitation. Biotechnology was coming to save them and William A. Haseltine, one of the few people in the world to live in the thin, pure air around the pinnacles of science, business, and politics, was about to tell them that they were not crazy, that just about everything they hoped for would someday come true. A few years after his meeting with Venter, Haseltine would lay out the full dimensions of his vision: "Practical immortality. That is my concept."

As the border between mainstream science and what was once considered the wacky fringe began to blur, a new religion was forming. Though neither side fully realized what was happening—and some denied it was happening at all—a most improbable cast of characters was helping to create the rapture. Devotees of psychedelic drugs, great thinkers, computer gurus, pioneering molecular biologists, a beautiful socialite, one of the world's greatest salesmen, science fiction buffs, and a renegade billionaire, all driven by a shared vision and by mutual enemies, would find themselves in common cause. It wasn't just everlasting life they were after, but *better* life, the bioutopia of sci-fi dreams. Indeed, in an echo of Father Wolf, the priest who taught my Catholic high school religion class, Haseltine himself would tell me that human beings would have "a transubstantiated future."

Oh, yes, the rapture was coming, alright. But whether it would look like the dreams of the bioutopians, or those who feared them, was as yet unknown.

2   THE PROPHET

THE IDEA THAT human beings could have a transubstantiated future might seem insane, or just deliberately provocative, a way for a guy like Haseltine—who is not only unafraid of controversy but seems to court it actively—to get his name in a magazine article or make a juicy sound bite for TV. But while Haseltine sounds radical now, he is only picking up where the mainstream left off before World War II, the aftermath of which changed the biological and social agendas. Though they have always been controversial, life extension and the enhancement of the human species were once regarded as noble causes. Indeed, it wasn't the fringe that originally inspired the science, it was the science that inspired what was to become the fringe.

On February 4, 1923, John Burdon Sanderson Haldane, a precocious thirty-one-year-old biologist, stood in front of Cambridge University's Heretics Society and presented a talk entitled "Daedalus or Science and the Future," a fantasy about the direction science, especially his field of biology, would take mankind in the next century. The Heretics Society might have seemed an odd forum. It had been organized more than a decade earlier to discuss religion, not science. But as the society's members understood all too well, and as Haldane reveled in telling them, people like Haldane were busy dismantling religion and the social order that supported it.

"The conservative has but little to fear from the man whose reason is the servant of his passions, but let him beware of him in whom reason has become the greatest and most terrible of the passions. These are the wreckers of outworn empires and civilisations, doubters, disintegrators, deicides. In the past they have been, in general, men like Voltaire, Bentham, Thales, Marx, and very possibly the divine Julius, but I think that Darwin furnishes an example of the same relentlessness of reason in the field of science. I suspect that as it becomes clear that at present reason not only has a freer play in science than elsewhere, but can produce as great effects on the world through science as through politics, philosophy, or literature, there will be more Darwins."

Haldane's celebration of Darwin and of reason as forces capable of overturning the accepted order had deep roots that ran all the way back to the Greek philosophers. In the 1200s, the English Franciscan monk Roger Bacon introduced the Church to the study of the natural sciences, the clear-eyed observation of phenomena as they happened, which were then reported with unvarnished truth. Through science, Bacon thought, one could learn truth and though his truth led to the Catholic God, the idea that scientific investigation could produce tangible benefits in the earthly world took hold. With the coming of the Renaissance and the Age of Reason, science came to be seen as the solution to many of man's ills. In 1626, Francis Bacon wrote his utopian vision of New Atlantis, an island called Bensalem which thrived thanks to a dedication to "the knowledge of causes, and secret motions of things, and the enlarging of the bounds of human empire, to the effecting of all things possible."

The technology this science created led to an array of useful salves, like caves used "for curing of some diseases, and for the prolongation of life" and water "which we call water of paradise, being by that we do to it made very sovereign for health and prolongation of life." Bensalem was equipped with "certain chambers which we call chambers of health, where we qualify the air as we think good and proper for the cure of divers diseases and preservation of health." The realm featured rejuvenation spas: "We have also fair and large baths, of several mixtures, for the cure of diseases, and the restoring of man's body from arefaction; and others for the confirming of it in strength of sinews, vital parts, and the very juice and substance of the body."

Bensalem was a biotechnology pioneer, manipulating the biology of animals. "By art likewise we make them greater or smaller than their kind is, and contrariwise dwarf them and stay their growth; we make them more fruitful and bearing than their kind is, and contrariwise barren and not generative. . . . Neither do we do this by chance, but we know beforehand of what matter and comixture, what kind of those creatures will arise." Biotechnology even allowed the Bensalemites to enhance human beings. "We have some meats also and bread, and drinks, which, taken by men, enable them to fast long after; and some other, that used make the very flesh of men's bodies sensibly more hard and tough and their strength far greater than otherwise it would be."

In the eighteenth century, French materialism proposed explanations for the mystery of man. Descartes's famous "I think, therefore I am" was a statement in support of the brain as the seat of whatever it meant to be a human being. Julien Offray de la Mettrie took this idea even further. In his 1748 book, *Machine Man,* he proposed that man was, literally, a biological machine that, as difficult as it would be, could eventually be fully understood by studying all of man's component parts. Differences and amounts of human brain was what really separated man from animals and, in a theory that would echo in the late 1990s in work done by Francis Crick who proposed that the "soul" was most probably a group of neurons, that what people regarded as the human spirit was actually a component of the brain. This idea of materialism—its theory that man was a collection of parts—and new experiments with electricity influenced Mary Shelley when she dreamed up *Frankenstein* in 1816.

If science could understand man, then science could adjust man. With the coming of the Industrial Revolution and its increasing control of technology, this did not seem like such a radical idea. Science became less of a philosophy and more of a way to get things done, and the people who worked in science became a class of intellectuals unto themselves, no longer called "natural philosophers" but, using a word invented in 1840 by astronomer William Whewell, "scientists."

The father of modern science fiction, Jules Verne, created his vivid novels out of this new thinking. Inspired by real innovations like the dynamo, Verne imagined miracles like sophisticated submarines and spaceships. Though science and technology could clearly be dangerous—Captain Nemo used the *Nautilus* for revenge—Verne presented science as a glorious enterprise.

Verne's English-language scion, H. G. Wells, in the February 6, 1902 issue of *Nature* summed up the excitement of the new twentieth century, celebrating the potential of science to create a new, better world, including improved generations of people:

> It is possible to believe that all the past is but the beginning of a beginning, and that all that is and has been is but the twilight of the dawn. It is possible to believe that all that the human mind has ever accomplished is but the dream before the awakening. We cannot see, there is no need for us to see, what this world will be like when the day has fully come. We are creatures of the twilight. But it is out of our race and lineage that minds will spring, that will reach back to us in our littleness to know us better than we know ourselves, and that will reach forward fearlessly to comprehend this future that defeats our eyes. All this world is heavy with the promise of greater things, and a day will come, one day in the unending succession of days, when beings, beings who are now latent in our thoughts and hidden in our loins, shall stand upon this earth as one stands upon a footstool, and shall laugh and reach out their hands amidst the stars.

Wells was predicting a coming rapture ("when the day has fully come") when future generations would constitute an improved species ("beings who are now latent in our thoughts and hidden in our loins") who would command nature ("stand upon this earth as one stands upon a footstool") rather than be commanded by it. Those future generations would exploit the "promise of greater things." Concepts like these were the seeds nurtured by later bioutopians calling themselves "transhumanists," a movement, it would turn out, that would play an integral role in trying to make Wells's prediction come true.

Wells, who studied biology under Thomas Huxley, Darwin's great defender, believed that science and politics were inseparable and that, as Haldane would later argue, religion had to be usurped by reason. He was a Fabian, a Socialist who supported the ongoing project of the perfectibility of man. Science, he thought, would play a large part in the great project, but only if it could manage to break the bonds of religion. His short 1896 novel, *The Island of Dr. Moreau*, often read as a warning about the dangers of biological tinkering, is, in fact, an allegory attacking religion and the social conventions of the time. Moreau sailed to his island because he "was simply howled out of the country. It may be he deserved to be, but I still think the tepid support of his fellow investigators, and his desertion by the great body of scientific workers, was a shameful thing." Cut loose from his place in society, left to his own devices on the island, Moreau turns himself into God of the beast men, who are themselves representations of mankind hobbled by tradition, ignorance, unreasoned faith. "Poor brutes! . . . Before they had been beasts, their instincts fitly adapted to their surroundings, and happy as living things may be. Now they stumbled in the shackles of humanity, lived in a fear that never died, fretted by a law they could not understand." The narrator of the story, who also studied under Huxley, fears for his own life after Moreau's death, and so creates the Jesus myth of Moreau. "'Children of the Law,' I said, 'he is not dead . . . He has changed his shape—he has changed his body,' I went on. 'For a time you will not see him. He is . . . there'—I pointed upward— 'where he can watch you. You cannot see him. But he can see you. Fear the law.'"

Very little separated man and beast. Biology was an accident. Reason and civilization make the man. That's what evolution taught Wells. Even when the narrator returns to London, rescued from the island, the words of a preacher sound to him like the gibberish of the ape man on the island.

Haldane was a product of this long tradition of science-as-salvation and of reason above all. He was born in 1892 during Victoria's reign, the son of a physiologist. He was educated at Eton and Oxford and then served in the famed Black Watch regiment in both France and Iraq during World War I. The war left him a changed man—as it did Wells, who toured the trenches—and radicalized him. Many other veterans also returned home embittered, rejecting for the first time values like blind patriotism and strict social mores. Poet Sigfried Sassoon epitomized the attitude of many soldiers:

The House is crammed: tier beyond tier they grin
And cackle at the Show, while prancing ranks
Of harlots shrill the chorus, drunk with din;
"We're sure the Kaiser loves the dear old Tanks!"

I'd like to see a Tank come down the stalls,
Lurching to ragtime tunes, or "Home Sweet Home,"—
And there'd be no more jokes in Music-halls
To mock the riddled corpses round Bapaume.
                                        —"Blighters," 1917

World War I was a spectacular example of how the old morality was unable to cope with the forces unleashed by science. It was the first chemical, mechanized war. The trenches filled with mustard gas. Men were ground up by machines. Haldane did not use the war as a rationale for halting science, but one for expanding it so that human beings could overcome the irrational passions that led to war. "The prospect of the next world-war has at least this satisfactory element," Haldane told the Heretics. "In the late war the most rabid nationalists were to be found well behind the front line. In the next war no one will be behind the front line. It will be brought home to all whom it may concern that war is a very dirty business."

Society changed quickly after the war. Royal dynasties like the Hapsburgs crumbled. Marxists took over Russia and communism seemed like a viable alternative to the kinds of governments that had fed the war. Haldane himself became a Communist. Feminism was born, and by the time Haldane was speaking to the Heretics, jazz was heard on new radio sets. As in jazz, improvisation was considered a virtue. "Free thinkers" and socialists banded together to sculpt a new culture out of the tattered one. Haldane's own circle of free thinkers included his sister, the writer and newspaper correspondent Naomi Mitchison; her husband, Dick Mitchison, a member of Parliament; the Huxleys, Julian and Aldous, Thomas Huxley's grandsons. As Haldane's former pupil Krishna Dronamraju points out in *Haldane's Daedalus Revisited*, these associations brought Haldane into contact with the intellectuals of Bloomsbury and their friends: Lytton Strachey, Virginia Woolf, T. S. Eliot, Yeats, Eddy Sackville-West.

Birth control came to America and Britain. D. H. Lawrence and William Butler Yeats scandalized society with their novels. Thinkers like Alfred North Whitehead, John Dewey, Bertrand Russell, all influenced by Darwinism, were creating new philosophical thought. Expressionism, which had begun before the war with works like Edvard Munch's *Scream*, in 1893, now exploded in Germany, Austria, and America with art by war veterans like Otto Dix, George Grosz, and Oskar Kokoschka. The cynical, hopeless moodiness depicted in the paintings found their way into the new cinema, most famously in *Nosferatu*, the precursor to Dracula movies.

Out of the postwar moral confusion, a new compass had to replace the old, thought Haldane, and it should be based, largely, on the ability of science to cut through confusion to get at truth. So to Haldane and those in the wide circle of which he was a part, experimentation—whether in marriage, art, sex, or politics—became a way of life. Reason, not superstition, not God, would serve as man's guide, and the highest expression of reason was science.

The year before Haldane spoke to the Heretics, Sir James George Frazer published a one-volume edition of *The Golden Bough*, his multivolume, monumental work on the ancient Greek priesthood of Diana, a study that came to encompass a panoply of ancient myths, religions, rites, and cultures, many of which had roots in early agriculture. Frazer came to the conclusion that religion was merely a comforting soporific in the face of death.

> Above all the thought of the seed buried in the earth in order to spring up new and higher life readily suggested a comparison with human destiny, and strengthened the hope that for man, too, the grave may be but the beginning of a better and happier existence in some brighter world unknown. . . . No doubt it is easy for us to discern the flimsiness of the logical foundation on which such high hopes were built. But drowning men clutch at straws, and we need not wonder that the Greeks, like ourselves, with death before them, and a great love of life in their hearts, should not have stopped to weigh with too nice a hand the arguments that told for and against the prospect of human immortality. The reasoning that satisfied Saint Paul and has brought comfort to untold thousands of sorrowing Christians, standing by the death bed or the open grave of their loved ones was good enough to pass muster with ancient pagans, when they too bowed their heads under the burden of grief, and with the taper of life burning low in the socket, looked forward into the darkness of the unknown.

But in the modern, post-Darwin age, what happened if you killed God? Where did that leave you? Staring into an empty hole in the ground? No. "Here at last," Frazer declared, "after groping about in the dark for countless ages, man has hit upon a clue to the labyrinth, a golden key that opens many locks in the treasury of nature. It is probably not too much to say that the hope of progress—moral and intellectual as well as material—in the future is bound up with the fortunes of science, and that every obstacle placed in the way of scientific discovery is a wrong to humanity." "In short," Frazer wrote, "religion, regarded as an explanation of nature, is replaced by science."

This view did not go entirely unchallenged. As the war made so obvious, science and technology carried dangers along with promise. Even when warning shots were fired, however, the message could still erode the way society viewed man's place in the universe, what it meant to be human, and reinforced the idea that man could usurp God. Fritz Lang's movie *Metropolis* showed humans serving machines, but also humans becoming interchangeable with machines. In their play *R.U.R.*, the Czech brothers Josef and Karel Capek introduced the world to "robots," machines made in the image and likeness of man to do man's bidding. But the robots, animated by a tiny amount of a substance called "biogen," acquire minds of their own and kill every human being except one, a clerk working for Rossum's Universal Robots named Alquist. The robots begin to die off because they cannot have children and they plead with Alquist to create more life. "Life will not proceed from test tubes," he tells them. In the end, however, the robots do become human. They develop emotions. Man, who created them, becomes God.

R.U.R. was first produced in London three months after Haldane's address. While Haldane agreed that man could, and should, assume control from God, he did not believe that the risks of taking a hand in human biology outweighed the benefits. He agreed with Frazer. Science would shape a new reality out of the husk of the old. "The time has gone by when a Huxley could believe that while science might indeed remould traditional mythology, traditional morals were impregnable and sacrosanct to it," he argued in *Daedalus*. "We must learn not to take traditional morals too seriously. And it is just because even the least dogmatic of religions tends to associate itself with some kind of unalterable moral tradition, that there can be no truce between science and religion." The mythical Daedalus invented flying and glue and defied Minos, the son of Zeus. And yet, Haldane argued, Daedalus appeared to suffer no eternal damnation.

"He was the first to demonstrate that the scientific worker is not concerned with gods." Men who follow their passion for reason, Haldane told the Heretics, will create a new biology. They will "throw down their dragon's teeth into the world." Human eggs would be cultured and fertilized in labs! Children would be born of artificial wombs. The very biological makeup of offspring would no longer be left to chance. Psychological states could be shaped by drugs. The "chemical substance produced by the ovary"—now called estrogen—would be synthesized and used to prolong a woman's youth and alleviate symptoms of menopause.

Daedalus was later published as a booklet and just as *Catcher in the Rye* is to disaffected English lit majors, so *Daedalus* was to imaginative biology students of the 1920s and the three generations since. (Four decades after *Daedalus*, Joshua Lederberg, one of the founders of modern molecular biology, declared that it, and Haldane's other popular writings, were "some of the liveliest and most intelligent prose on scientific subjects that has ever been written.") Haldane is a hero not only because of his prescience, his uncanny predictions of in vitro fertilization (IVF), gene therapy, and hormone therapy, but because he gave biology power. The speech challenged young biologists to throw off old presumptions, to aim high. Biology, Haldane argued, could be made to bend in the direction of man's choice. Do not just describe phenomena, he told them, *do* something about it. Haldane made biology sexy.

But *Daedalus* also incited an uproar, both because of Haldane's provocative statements on social mores and because of his predictions about the "dragon's teeth" he imagined biology would throw into the world. Haldane's vision so worried his friend Aldous Huxley that Huxley used elements of it in the dystopic satire *Brave New World*. And just as Haldane's vision would be a rallying cry for future bioutopians, so the warning in Huxley's novel would become a handy cliché for generations of people opposed to the manipulation of human biology, foreshadowing a conflict that would erupt anew at the end of the century.

THE PROBLEM, as Haldane knew, was that while man could imagine taking control, there were not many "dragon's teeth" to throw in 1923. "We are almost completely ignorant of biology," he told the

Heretics. As proof, Haldane pointed to the one attempt so far to take a hand in evolution, and there science botched the job.

Following Darwin's publication of *Origin of Species* in 1859, science almost immediately tried to apply the principles of natural selection to human beings, an endeavor seen in many quarters of society as a progressive impulse. Darwin's cousin, Francis Galton, studied the effect of natural selection on the accomplished men of Britain and discovered, not surprisingly, that talent tended to run in families. Galton downplayed the effect of nurture and environment and focused on biological inheritance. Nobody knew how such inheritance worked, what the heritable mechanism was, but it was obvious to Galton that something transmitted the admirable qualities he found in luminaries to their offspring.

Galton came to believe that with appropriate tweaking man's nature could be shaped to achieve desirable ends, a betterment in the general welfare. In 1883, he invented a new word to describe a system to create an improved human species, *eugenics*, a hybrid of two Greek words meaning "well born." If the most talented people would seek each other out to reproduce, and if the least talented— the defective, the ignorant—would be discouraged from having children, the entire species would improve.

Today, of course, eugenics is a word so burdened with the history of American race laws and Nazi extermination camps that Galton's progressive purpose is completely overshadowed. But he wanted to improve the lot of man—not just of the upper classes, but of the poor as well. His message found a ready audience among both progressives and conservatives who were increasingly alarmed at what they saw as an unbalanced birth rate between the talented and the ignorant, a discrepancy that, they worried, would eventually lead to the destruction of the country and race suicide.

Galton's ideas became so popular that the American socialist Edward Bellamy adopted them for his own utopian vision. In *Looking Backward*, Bellamy's narrator falls asleep in 1887 and wakes up in Boston in 2000, to a society shaped by eugenics. Crime is understood to be atavistic, the narrator is told, "because nearly all forms of crime known to you are motiveless now, and when they appear can only be explained as the outcropping of ancestral traits." By being choosy about mating, "every generation is sifted through a little finer than the last. The attributes that human nature admires are preserved, those that repel it are left behind."

Back in the real world, eugenic laws were passed and eugenic societies formed in Europe and the Americas. The seeds of eugenics found especially fertile ground when they blew into the United States. Theodore Roosevelt embraced it, a center to promulgate eugenics was established at a somewhat ramshackle Long Island laboratory called Cold Spring Harbor, and, by the 1920s, state and county fairs ran eugenics competitions, awarding prizes to "Fitter Families." Before the eugenics craze ended in America, tens of thousands of people had been ordered sterilized.

Haldane believed eugenics was silly, a false start on the road to human betterment and an example of how little was yet understood about human biology, though it did have the beneficial effect, he told the Heretics, of preparing "public opinion for what was to come." Haldane had an idea of what was coming because he knew that some of the most basic workings of biology were just beginning to show themselves.

The idea of the gene, for example, had a sixty-year history and was already well accepted. At about the same time Darwin was roiling both science and religion with *Origin of Species*, an Austrian monk, Gregor Mendel, was in the midst of experimenting with pea plants to deduce how traits like color are passed on from one generation to the next. He concluded that traits were handed from parent to offspring by packets of some hereditary material.

Mendel's work went almost unnoticed for the next forty years, until the turn of the century, when botanists rediscovered his writings. One, a Dutchman named Hugo de Vries, expanded upon Mendel's work and dubbed Mendel's hereditary units *pangenes*, a union of the Greek words for "all" and "stem." Despite naming them, however, nobody, including de Vries, was sure such hypothetical units existed.

Before the turn of the twentieth century, scientists relied mostly on a few shekels from a college or on family money to pay their way in a life of research, putting large-scale experiments out of reach for many. But the Industrial Revolution not only provided railroads, the telegraph, and new electric systems, it also created great fortunes. The men who controlled this wealth, flush with the optimism of thinkers like H. G. Wells, and wanting to improve mankind as they had improved machines, decided it was time to create a Biological Revolution that could have the same practical impact on man as the industrial one had on machines. In 1892, for example, wealthy Philadelphia lawyer Isaac Jones Wistar provided funds to expand a specimen museum originally created by his great-uncle, and to turn it into the Wistar Institute of Anatomy and Biology. By 1906, the institute had created the world's first standardized lab animal, the WISTRAT. About half of all lab rats are descended from this original line. In 1901, John D. Rockefeller used part of his Standard Oil fortune to establish the Rockfeller Institute for Medical Research (now Rockefeller University) in New York City.

While private money funded new institutes in the United States, the government took the lead in Britain. In 1911, it founded the Medical Research Committee, later renamed the Medical Research Council (MRC), to coordinate and fund a national research effort. This was the very beginning of what is now known as "big science."

Supplied with this new money, biologists intensified research into genes as hereditary units. A Rockefeller researcher, Thomas Hunt Morgan, selectively bred fruit flies to create mutations. His work proved that whatever the genetic information was, it was carried on the "colored bodies" inside the cell, the chromosomes.

Haldane was also aware of the ways in which biology had already been harnessed for useful purposes. In Britain, an agricultural scientist named John Hammond decided that science in labs, as interesting as it was, wasn't much good if you couldn't see real results you could use. So he married basic science to practical application decades before most other scientists dreamed of such things. "I wanted to apply scientific work to growth in farm animals," he told an interviewer in 1962. "But I couldn't find any work of this kind, and so I had to do some myself. All through, when I wanted to know something that wasn't in the scientific literature, I have had to go into pure science myself to dig out the facts so that I could apply them." Hammond, who entered Cambridge in 1906 and began serious lab work in the 1910s, did not place much faith in theoretical statistics to predict the outcome of experiments. What would you get when you crossed a tiny Shetland pony mare with a giant Shire stallion? Well, let's find out. What happens when you inject growth hormone into pigs? Let's try it. Hammond's attitude, and his history, show why some of the most important developments that would come to influence the thinking of scientists like Haseltine, and the thinking of the bioutopians, would arrive from the obscure left field of farming. These developments would include IVF, the world's first clone of a mammal, and the growth of human embryonic stem cells in dishes.

After serving in the army during World War I, Hammond came home to England determined to improve livestock for a hungry nation. Since some experiments in the artificial insemination of horses had already been tried, Hammond decided to see if he could quickly inseminate a large number of cows using semen from the best bulls he could find. If it worked, the country could ramp up milk production and possibly prevent the diseases that were sometimes transmitted when farmers brought together cows and bulls from different farms. But Hammond had no large animal facility after the war. He had to settle for inseminating rabbits as a test. In 1922, the year before Haldane presented "Daedalus" to the Heretics and predicted IVF, Hammond and a colleague sent some rabbit semen to Edinburgh, where it was shot into females. The process worked. Live, healthy rabbits were born, the first time that mammal babies were made without sex.

Hammond then sent ram semen to Poland and workers there got baby sheep. He demonstrated the process to a visiting scientist from Denmark who went home, used it in cattle, and got calves. But despite the success elsewhere, Hammond met with religious opposition when he tried to use it in England.

"I met the Committee of the Church Council, I think it was, and put the case. They attacked me first by saying that it was artificial, and nothing artificial could be of any use. It was going against nature. . . . The second objection was on moral grounds." The clerics feared that artificial insemination might be used on people. "I said, 'I am not going to argue human, I am going to argue cattle, which we want it for. Humans are not my line.'" To overcome the argument that the technology was inherently immoral, separating reproduction from sex, and therefore corrupting, Hammond reminded the church-

men of how cattle breeding worked at the time. In villages throughout England, he said, farmers paraded bulls through the middle of
town to visit the cows of other farmers. "Everyone knows what is happening and all the small boys come and watch," he told them. How
could a farmer with a syringe be more corrupting than a farm animal
sex show? he asked. The Church eventually acquiesced, but only,
Hammond said, because of the intervention of the minister for agriculture during the Second World War.

There were other developments that inspired Haldane, especially
in the field of developmental biology, the study of how animals are
built from the embryo up. Thinkers had been noodling over this mystery for centuries. Aristotle thought sperm were the seeds of people
and that men sowed the seeds in the fertile acreage of the womb. By
the late 1600s, many scientists thought people existed inside the
sperm as tiny but complete forms called homunculi. Woodcuts from
the early 1700s show an intrepid submariner packed inside his
spermy vessel.

In the eighteenth century, biology discovered a neat party trick.
If you cut up certain animals, the pieces made new ones. One such
animal, a hydra, was a glob of cells that commonly lives in freshwater ponds. Another was a tiny flatworm called a planaria. Both the
hydra and the planaria possessed total powers of regeneration. Slice
a planaria in half and you'd wind up with two planaria.

During the late 1890s, Thomas Hunt Morgan, among others,
diced planaria up into small pieces and, sure enough, most of the
pieces regenerated entirely new planaria. A chunk of tail could be
sliced away, and it would start making a whole new worm, including
worm brains, worm muscles, worm mouth parts. There were no
brains in the chunk of tail, no mouth parts. But the cells in the tail
knew how to make them anyway. Scientists also knew that if you
chopped the leg off a newt, the newt would grow a new leg. Some
lizards would grow new tails. Clearly, the cell had powers as yet
unfathomed.

During the first decades of the twentieth century, scientists
turned their attention to mammals. In the course of studying how
mammals develop, they sometimes induced parthenogenesis, the
division of unfertilized eggs, with chemicals and electricity. Often
the "parthenotes" formed early embryos despite the fact that the egg

had never met a sperm. The egg just did it on its own. This work sparked brief excitement when, in 1911, an article on parthenogenesis appeared in the original *Scientific American* headlined "The Creation of 'Artificial Life.'" (In the 1950s, J.B.S. Haldane's wife, biologist Helen Spurway, thought parthenogenesis could have created a living human being, a virgin birth like that foretold by Isaiah: "Behold, a virgin shall conceive and bear a son and call his name Immanuel." Some higher animals were known to have reproduced this way, including turkeys, sharks, and snakes. So Spurway mounted an effort to find a parthenogenic human. She advertised for candidates, most of whom misunderstood and thought she was referring to bastards, but finally narrowed the field down to several possible candidates. In the end, she rejected them all.)

But Haldane realized that the business end of human change would center on genes, and as a geneticist, he knew that while science accepted their existence, they were still largely mysterious. Nobody knew exactly what relationship they had to the chromosomes, how they were copied, and what passed them on to the next generation. Still, Haldane speculated that such knowledge would be mastered soon and that life would someday be made by design, just as Francis Bacon had said 300 years earlier.

Just three years after Haldane's speech, Thomas Hunt Morgan's colleague at Rockefeller, Hermann Muller, created the first artificial mutations in animals by subjecting fruit flies to X rays. For the first time, an artificial force, an X-ray machine, had changed the basic structure of an animal. This was a very tangible, practical impact, not a trick, not a change created by the old-fashioned method of breeding as Morgan had done. The media took up the Muller experiments and said that someday, maybe soon, scientists would create designer people.

The public began to take notice of biology not just because of Muller's experiments but also because in 1926, Hugo Gernsback founded the first pulp science fiction magazine, *Amazing Stories*. Mutation via X rays became a sci-fi staple and genes entered the common parlance. Fictional biologists made frequent appearances in *Amazing Stories*, which Gernsback saw as a vehicle to introduce lay readers to the inside world of science. But the new pulp science fiction could not match the coming reality. Soon, the world would wake up to just how powerful science could be. In the aftermath of that white light shock of the atomic mushroom cloud, biology would take a step back from Haldane's future. Yet even as biology's public outlook would become somewhat more conservative, it was accumulating all the tools it would need to call Haldane's prophecy back to life.

3   THE ENDLESS FRONTIER

DURING THE WANING DAYS of World War II, the U.S. Office of Scientific Research and Development issued a report entitled "Science: The Endless Frontier." It was the product, mainly, of the office's director, Vannevar Bush. Bush, a Harvard and MIT man famous for creating a precursor to the modern computer, and who would go on to become chairman of the pharmaceutical giant Merck, thought a strong national investment in science was critical for the nation's future security and economic well-being. In response, Congress created the National Science Foundation (NSF) in 1950. It also began generously funding the NIH, supplying it with money to dole out to researchers around the country. Scientists soaked it up and the pursuit of basic research funded by government became ingrained in university, military, and public health labs all over America.

This made President Dwight Eisenhower nervous. "In this revolution research has become central; it also becomes more formalized, complex, and costly," he told the nation in his farewell address on January 17, 1961. "A steadily increasing share is conducted for, by, or at the direction of the Federal government."

Eisenhower was mainly concerned about military research and the creation of a "military-industrial complex." But he also had in mind science in general, including the biological sciences.

"Crises there will continue to be. In meeting them, whether foreign or domestic, great or small, there is a recurring temptation to feel that some spectacular and costly action could become the miraculous solution to all current difficulties. A huge increase in newer elements of our defense; development of unrealistic programs to cure every ill in agriculture; a dramatic expansion in basic and applied research—these and many other possibilities, each promising in itself, may be suggested as the only way to the road we wish to travel." Don't you believe it, Eisenhower warned.

But the Allies had won World War II, and they did it in no small part thanks to scientists. There was the A-bomb, of course, but also the code crackers, penicillin, and radar. Everyone knew the consequences of allowing the national technology effort to fall behind an adversary's. They had seen the results of German V-2 rockets falling on Britain and they had heard the stories of American air crews lumbering along in propeller-driven bombers as the first enemy jet aircraft whizzed by during the last days of the war. Now the Iron Curtain had descended upon Europe and America faced a new enemy, the Soviet Union. Science and scientists became the West's most important weapon.

William Haseltine was born into this new world, at its very core, in fact, and while he is not a sentimental guy, his eyes grow almost misty at the memory of his childhood. But as he sits in the living room of the weekend New York pied-à-terre he shares with his wife, Gale Hayman, and talks about growing up during the late 1940s and early 1950s in a place a world away from the one he occupies now, it's clear that he's editing. He is creating his own mental movie, a very long episode of the old "Wonderful World of Disney" in which crew-cut kids in dungarees and flannel shirts and baseball caps made walking sticks out of tree branches and sauntered forth on exciting explorations. That was true, as far as it goes. While Haseltine's youth was not idyllic, it did instill a drive and ambition that would become legendary among those who would later work with him or compete against him. And it carved a permanent desire not just to be thought great, but to actually *be* great, an aspiration that helped motivate him to declare human immortality an achievable goal and to regard himself as one of the world's Haldanes who could supply a vision of the human future.

Haseltine was born into science. His father, William R. Haseltine, grew up in Ripon, Wisconsin, a town in Fond du Lac County not far from Oshkosh that was first settled by an early, proto-socialist utopian movement called the Phalanx. The Haseltines had deep roots there. Charles Haseltine came to Ripon in the mid-1800s and, in 1859, firmly established the family presence by purchasing an impressive-looking brick house with an old, white, one-room school building in the back yard. That Little White Schoolhouse, as it became known, was the site of the first organizing meeting of the Republican Party on February 28, 1854 (though Michigan also claims this honor). Today, the property is owned by Bing and Khien Lam, Vietnamese immigrants, who have turned the place into the Republican House Restaurant. They serve "American/Chinese Cuisine Dine In and Carryout." There's a mural featuring Abraham Lincoln and Theodore Roosevelt over the bar.

William R. attended Ripon College, then the University of Wisconsin, and finally retraced his father's footsteps to MIT, where he earned a Ph.D. in physics. He accepted a postdoctoral position at the University of California at Berkeley, where he met and married Jean Adele Ellsberg.

When World War II broke out, William R. did what a son of Ripon and the son of a veteran who was an active supporter of Reserved Officers Training Corps (ROTC) did. He joined the army. There was certainly a more interesting place for a brilliant physicist with a Ph.D. from MIT out in the high desert of New Mexico working on the Manhattan Project, but William R. did not wait to be recruited. He entered an army training program and emerged as a logistics officer assigned to Philadelphia.

Not long after he and Jean moved to the base at Philadelphia, Jean gave birth to a baby girl, Florence. William R. had been hoping for a boy so Florence, whose middle name was "Pat" after a grandmother, quickly became "Paddy." When Florence was ten months old, the family moved on to the next posting, Jefferson Barracks in St. Louis, Missouri. Baby Bill was born at the base hospital on October 17, 1944.

The timing of his arrival into the world was improbably fortuitous for a child whose later professional life would revolve around genes. Pioneers were rapidly paving the road he would travel. Some would even become his mentors.

JUST THREE YEARS before Haseltine's birth, scientists George Beadle and Edward Tatum had proved that genes were responsible for the production of proteins, the workhorses of biology. That left a great unsolved mystery. What were genes made of? This became the most tantalizing problem in all of biology. "You may have heard that we are going to Stanford next month," Norman Horowitz wrote to fellow geneticist Max Delbruck on June 9, 1942. "I am going to work with Beadle on the problem of the future . . . the chemistry of the gene."

Chemistry was the right word, for now the realization was seeping into the minds of a few scientists that biology alone would not provide the tools needed to understand the processes of life. Delbruck himself was trained as a physicist, not a biologist, and was one of a handful of physicists who had become fascinated with biology.

Austrian physicist Erwin Schrodinger became an evangelist for this nascent movement, doing for the physics community what Haldane had done twenty years earlier for biologists. Schrodinger, famous for his role in developing the theory of quantum mechanics, winner of the Nobel Prize in 1933, and author of a 1935 quantum theory "thought experiment" involving a cat which may or may not be killed inside a box believed that the time had come to merge the sciences. In a February 1943 lecture at Trinity College, Dublin, where he lived in exile following the rise of the Nazis, Schrodinger told his audience that "it has become next to impossible for a single mind fully to command more than a small, specialized portion of" science. The solution, he thought, was for the sciences to cross-pollinate. "I can see no other escape from this dilemma (lest our true aim be lost forever) than that some of us should venture to embark on a synthesis of facts and theories."

Like *Daedalus*, Schrodinger's 1943 lecture was published as a book in 1944, the year of Haseltine's birth. The title was simple but profound: *What Is Life?* Schrodinger gave physicists much the same mission as Haldane imagined. Life, Schrodinger said, could be understood as a mechanism, something like a clock. "Few words more are needed to disclose the point of resemblance between a clockwork and an organism," he said, sounding a bit like la Mettrie. "It is simply and solely that the latter also hinges upon a solid—the aperiodic crystal forming the hereditary substance, largely withdrawn from the disorder of heat and motion." In other words, the heart of life was somehow sequestered from the cyclical processes of assembly and decay. It was not ephemeral but eternal. And it was not alive. It was a thing, like a crystal. Whatever the hereditary substance was—Schrodinger used the crystal as a metaphor—it worked according to physics. Change just a few atoms within the hereditary substance that was contained in a sperm or an egg, and you could change the species. There was a lot going on in physics in 1944, including atomic weapons research and speculation about the nature of the universe, but to Schrodinger "these facts are easily the most interesting that science has revealed in our day."

A lot of scientists thought so, too. And no wonder. The implications of Schrodinger's argument, which he spelled out in an epilogue, were enormous. Human beings were a mechanism. Yet human beings could control the mechanism. "It is daring to give this conclusion the simple wording that it requires. In Christian terminology, to say: 'Hence I am God Almighty' sounds both blasphemous and lunatic." Nevertheless, the ability to exert such control could "be condensed in the phrase: *Deus factus sum* (I have become God)."

Haldane, who was by then an eminence, read and was highly influenced by *What Is Life?* So were American chemist Linus Pauling and a young English physics student named Francis Crick. Students at the other centers where physics and biology could mix, like the California Institute of Technology in Pasadena where Pauling worked, Rockefeller, the University of Wisconsin at Madison, the University of Illinois at Urbana-Champaign, Chicago, Indiana, Michigan became excited about using physics to decipher life itself.

Maurice Fox was among these students. He had studied physics as an undergraduate at the University of Chicago, but transitioned into biology as a graduate student. His wife, Sally, recalled just how influential Schrodinger's call to arms was. "When we were married, Maury handed me a little book called *What Is Life?* by Schrodinger," she said. "He said, 'Here, read this. This is what I do.'"

"There weren't any other books!" Maury said, laughing.

"So I read it," Sally continued, "and after I finished it, I turned to him and I said, 'Why?' and he said, 'You are not allowed to do that! You can't ask why! You can ask how, when, and where, but never why!'"

"Why" was irrelevant. "Why?" could never be rationally answered, so what was the point? But the other questions, the "how" and the "when" and the "where," now *those* were questions science could, and was about to, answer. Schrodinger did not know it, but as he was addressing Trinity College, a Rockefeller scientist, Oswald Avery, was discovering the molecule of heredity.

Avery was approaching retirement at the time and was considering leaving the institution. But he had spent the past fifteen years obsessed with a question first posed by a British scientist in 1928 and he did not want to leave any unfinished business behind.

The question was a puzzle about two types of bacteria. One type was deadly, but it had been killed and so rendered impotent. A second type was alive, but harmless. But when both types were injected into mice, some of the mice died from infection. Somehow, the killed bacteria had transformed the harmless bacteria into a virulent form. At first, Avery did not believe those results, but he soon duplicated them in his own lab, even without the use of mice, simply by combining both types of bacteria in lab glassware.

"For the past two years, first with [Colin] MacLeod and now with Dr. [Maclyn] McCarty I have been trying to find out what is the chemical nature of the substance in the bacterial extract which induces this specific change," Avery wrote to his brother in May 1943. "Some job, full of headaches and heartbreaks! But at last, *perhaps*, we have it!" After repeated extractions and purifications, Avery wrote, "there separates out a fibrous substance which on stirring the mixture wraps itself about the glass rod like thread on a spool and the other impurities stay behind as granular precipitate. The fibrous material is redissolved and the process repeated several times. In short, this substance is highly reactive and on elementary analysis conforms *very* closely to the theoretical values of pure desoxyribose nucleic acid (thymus) type. (Who could have guessed it)."

Oswald Avery had discovered that DNA, a chemical molecule, was the "transforming principle," the key to heredity. "If we are right," Avery told his brother—and Avery seems to have barely believed it himself—"then it means that nucleic acids are not merely structurally important but functionally active substances in determining the biochemical activities and specific characteristics of cells and that by means of a known chemical substance, it is possible to induce *predictable and hereditary* changes in cells." Making changes with X rays as Muller had done was fascinating, yes, but those changes were random and unpredictable. One fly might develop a goofy wing, while another fly from the same experiment might wind up with a new eye color. With the DNA molecule in hand, however, one could reasonably begin to think about lifting the hood of biology and tinkering with it. From the very beginning of the DNA era, then, scientists thought about using access to the transforming principle to engineer life.

Avery published his discovery in the *Journal of Experimental Medicine* on February 1, 1944. It was not, as a layman might expect, entitled "Eureka! Secret of Life on Earth Revealed! Genes Are Made of DNA!" The tribal code of science, especially in the 1940s, required researchers to stick strictly to the facts and hope that colleagues could see through the dense lingo and make a fuss. So the paper by Avery, McCarty, and MacLeod was given the impenetrable headline: "Studies on the chemical nature of the substance inducing transformation of pneumococcal types. Induction of transformation by a desoxyribonucleic acid fraction isolated from pneumococcus type III."

Dating the start of a movement is an invitation to an argument. But the year of Haseltine's birth, 1944, is as good a date as any to designate the start of molecular biology, the reductionist era, when scientists began to think that by understanding the tiniest bits inside the nucleus of the cell, it would be possible to understand and manipulate life itself. The handful of scientists working in the field—and it was just a handful—were fired by Schrodinger's challenge. The chemical that transmitted life's instructions had been found at last. The answers to the most important questions in biology, thought the scientists, would now bubble up from the mysterious soup. That's just what happened.

WILLIAM R. HAD MISSED OUT on his chance to build bombs during the war, but soon after he took a job with the U.S. Navy applying his physics skills as a civilian scientist at China Lake Naval Weapons Center in the California desert. There, he would become one of the world's foremost experts on ballistics, contribute to both the Sidewinder and Polaris missile programs, and provide an attractive target for rebellious children two decades later.

China Lake, named for an ancient, long-dry lake bed in what is known as the Indian Wells Valley, is unlike any other place in the United States, with the possible exception of Los Alamos. It squats between Death Valley to the east and the foothills of the Sierra

Nevada Mountains to the west. Los Angeles is about ninety minutes' drive to the south. Even today, China Lake appears forbidding. It's bone-dry for most of the year, a furnace in the summer, and bone-chilling on winter nights. The reservation has the unmistakable look and feel of a military outpost, and since its nearest neighbors are Fort Irwin National Training Center and Edwards Air Force Base, the entire desert region seems like an enormous, slapped-together boomtown for people in uniform.

But despite its appearance, China Lake was a privileged place in the 1950s, an island of wonder, especially for children, to whose eyes the base looked a lot like their own Endless Frontier. China Lake is not far from the big trees of the Sequoia National Forest, and the desert itself has attractions that appeal to adventurous youngsters like the Haseltine children, whose numbers expanded to four after the births of Eric and Susan. Bill enjoyed Cub Scouts and Boy Scouts. There were horses to ride and wide streets to bicycle in and good schools and nobody locked their doors. Best of all, a child could go feral.

"We were totally free," Haseltine recalled. "We were on this big military base with a big fence and a 20 mile-per-hour speed limit and armed guards. What does a kid want but to be safe? And I could literally walk five minutes outside my back door and be in nature, in the desert. From the time I was seven or eight, two or three of us would go into the desert and hunt lizards with our BB guns and bows and arrows. And we'd go hike up this little hill of volcanic rock and take a lunch and explore. It was fabulous."

His sister, Florence, agreed with Haseltine's memory of the natural world around China Lake, but their interior world was another matter. Bill was physically awkward. His awkwardness was aggravated by health problems. A bout of tonsillitis led to a tonsillectomy, appendicitis to an appendectomy, and, when he was eight, he contracted pericarditis, an inflammation of the pericardium, the lining that surrounds the heart. He was bedridden for several months. Perhaps because of these setbacks, or perhaps because of his own genetic inheritance, it was obvious that he was not the most coordinated kid on the base. As a youth, he applied the scientific method to determine why some kids excelled in sports and he did not. "A new game arrived when I was about in the sixth grade. It involved a rubber ball, like baseball, and I thought to myself, Okay, let's take a look to see if the kids who are good in other sports are good at this one too and sure enough, the kids who were good at other sports were good at this one. It's tough when people put the emphasis on sports that require good hand-eye coordination." He was a decent swimmer and, he recalls, a fast runner, so he was not completely left out of the traditional young male hierarchy. And though team sports mostly eluded him, he was given a letter for football in junior high school, an award he admitted was "generous."

Fortunately for Haseltine, he was growing up in China Lake, a place where children could impress each other and adults in ways that had nothing to do with sports.

For all its Disney-like attractions, China Lake was a pressurized academic community. Parents bragged about their children's grades, or, conversely, disguised a child's poor performance. This was, after all, a town devoted to science. With scientists gathered from the great academic technology centers like Stanford and MIT and Berkeley and Chicago, China Lake belied its dusty boomtown appearance to become a multidisciplinary commune. Many of the civilian wives were college women like Jean Haseltine, who had studied at Reed and then Cal. She spoke French, sometimes working as a French teacher. China Lake had its own branch of the American Association of University Women. Art, literature, politics, and theater were common topics of party conversation.

China Lake was full of Korean War flying aces, renowned scientists, high-ranking military officers. They were Haseltine family friends. Future astronauts Wally Schirra and Scott Carpenter were neighbors. So at China Lake, excellence, academic and otherwise, was both rewarded and expected.

But as youngsters, Florence and Bill struggled in school. Florence was dyslexic and excelled only in math. Bill was merely an average, sometimes a little below average, student as a young boy. William R. was not impressed. Even before the war, when he was teaching bright would-be physicists at Berkeley, he was a tough grader, and he was no less demanding of his own children. And Jean, recalled Florence, was embarrassed by her children's seeming inability to compete.

If they wanted to please their father, the children knew the way to do it was through schoolwork and especially science, his only religion

according to Florence. When they asked for help on difficult homework assignments or general questions about space or physics or
mathematics, William R. would spend hours working out problems
and explaining concepts. But when Florence asked about sex, he
handed her a copy of the *Kinsey Report* and told her to read it.

He was also a harsh disciplinarian who favored corporal punishment. Bill often took his swats stoically. He stood still, trying not to
show weakness, but Florence would dash round the room and, as a
result, would wind up taking more blows. She still carries a scar
from a horsewhip.

Being smart, smarter than anybody, the Haseltine children knew,
was the one way they could earn their father's respect and alleviate
some of the anxiety their mother felt. So they devoted themselves to
academics and gradually began to succeed, often through sheer
force of will. Bill Haseltine would later tell a Harvard colleague that
it was at China Lake that he learned not to ask for something he
wanted, then hope it would come his way, but to set his sights on it,
and then do whatever it took in order to obtain it.

Florence also developed strategies. She learned to compensate
for her dyslexia and eventually won a Westinghouse Science Fair
award and even a Betty Crocker Homemaking Award. In 1957, when
the International Geophysical Year was declared (later extended into
1958), Florence joined the astronomy club. *Sputnik,* the first man-
made satellite, was launched by the Soviet Union on October 4,
1957, and Florence and the other members of the club would rise in
the predawn morning, bicycle to a viewing spot called "the fence"
away from houses, and try to see the passing satellite through a tele-
scope. Soon after *Sputnik III* was launched on May 12, 1958, Florence spotted it. She was the first person in the United States to see
it. She was awarded a pin saying "International Geophysical Year."

Bill was inspired by his science teachers in junior high and high
school. They encouraged him to work in the base laboratories, where
he helped real scientists work with chemicals and conduct physics
experiments. Any shyness about his physical awkwardness faded. He
didn't need sports. In other communities, the young bravos may have
strutted with letters on their chests, but at China Lake, the young
bravos around Bill Haseltine strutted with slide rules. Academics
was their contact sport. "My peer group in high school was particu-

larly rich in bright students. I would say about ten of us who knew each other from the time we were very young were sort of more or less on par with one another and we competed for everything, whether it was English or French or whatever. I was never in an atmosphere were my peers were as generally bright until I was a graduate student at Harvard."

With his academic performance up, all Haseltine needed was confidence. His mother helped him build it. During his sophomore year, the principal chose four students to compete in a speech event. The topic was "Why I Want to Go to Japan." The winner of the speech contest would win a summer-long trip to Japan aboard a U.S. Navy ship. "My mother gave me some very good advice. She said, 'Don't tell them why you want to go to Japan. These are navy guys. They fought the Japanese. Tell them why you want the experience of being on a navy vessel.'" Haseltine won the trip. In all, thirty students from the western United States spent the summer aboard the troop transport. "We got to work all parts of the ship and at night consort with the officers and their daughters. That was terrific!"

The following summer, he performed well enough on an exam and in another speech contest to win a cross-country trip as part of an Odd Fellows and Rebeccas Pilgrimage for Youth. He and about two dozen other students crammed into a bus, stayed in churches, ate in local cafeterias or in homes. A year later, during the summer after his senior year, he took a slow boat to Europe and then bicycled around it.

Until these trips, Haseltine had very little idea of the world beyond the desert apart from what he gleaned from books. When he speaks of them, his face lights up—something it rarely does—with the memories. "I got to mix with the sailors on the boat!" he says with the tone of an excited teenage boy. More important, he learned that the world was a much more complex place than life on the navy base or his father's Republican politics had led him to believe.

The successes of the Haseltine children pleased their father, but failed to ease their mother's pain. They grew up watching Jean being tormented by one medical problem after another. Detached retinas left her temporarily blinded. Then came septicemia, a bacterial infection of the blood. "Septicemia is a terrible thing," Haseltine recalls. "You can actually watch it progress. There was a red streak that would go up her arm." Haseltine and his mother were close and her illness was a horrible revelation. "I was told that if it ever got to her heart, she would die, whatever that means, and so every day I would go in [to her room] and see what it looked like. You could actually measure it, see if it went up or down. It was a terrifying thing to see that. All she could do was soak in Epsom salts and even as a kid, that did not seem to me to be very effective."

Worst of all, Jean was afflicted with manic-depression. When Florence was eleven and Bill was nine, Jean suffered the first of several breakdowns. One day she locked herself in the bathroom. Florence tried to get in, but the door was locked. She looked through a peephole and saw red. Blood was everywhere. Jean had just slashed her wrists and her neck and was quickly bleeding to death. Florence ran to her father. He broke down the bathroom door and found Jean laying in a pool of blood. She survived, but over the next several years she would suffer periodic bouts of mental illness requiring admission to a hospital. In 1981, Jean Haseltine finally succeeded in taking her own life.

Nobody, Bill thought, should have to go through what his mother had gone through. He vowed to make sure. "I did not, even at a very young age, like what had happened to me," he says, referring to his own illnesses. "And I certainly did not like what had happened to my mother." So from the time he was a youngster, Bill knew what his life's mission would be. Though he briefly flirted with becoming a test pilot, he hardly wavered from the road he imagined walking from the time he felt the dread of stepping into his mother's room to chart the course of her septicemia. By the time he started high school, he was certain. "I was very specific. I wanted to be a physician."

But Bill Haseltine had something other than being a family doctor in mind. He wanted to change medicine. This was not an unreasonable ambition for a budding science geek in the 1950s, because biology was finding its legs, already starting to manipulate the simplest forms of life.

JUST AFTER THE WAR, Max Delbruck helped to create something called "The Phage Course," a weeks-long tutorial on bacteriophages, the viruses that infect bacteria, at the now-transformed eugenics facility, Cold Spring Harbor on Long Island. When phages infect bacteria, they bring along bits of their own genetic code, leaving some in the bacterial cell and picking up genetic instructions from the bacteria. Then, when the phage copies itself and infects a second bacteria, it can exchange these codes, thus transmuting one kind of bacteria into another. Because of this ability, the bacteria-and-phage system was the simplest model to use as a genetics study aid. Later, the courses expanded to bacterial genetics and animal viruses.

Cold Spring Harbor became a kind of summer camp and salon. Physicists and cell biologists and geneticists would set up house-keeping in primitive, damp cabins. ("If you brought new shoes, they would be moldy within two weeks!" Sally Fox remembered.) The scientists would bring along the wife and the kids and give free rein to their imaginations. Most were somewhat like the free thinkers of the 1920s, politically liberal, unconventional, areligious. All were conscious of the fact that they were now diving deep into the waters of life, deeper than anybody had gone before, and that what they discovered would someday change history.

Every summer seemed to bring some new excitement. "One evening they showed time-lapse cinematography," Sally recalled. A Polish cell scientist had managed to film cell division, mitosis, in corn. "Well, the chromosomes would all line up and then, as if by a command, the chromosomes would split! I would say there were thirty people in the audience, scientists and their wives—the only woman scientist was Barbara McKlintock—and this was Max Delbruck, Salvador Luria, Al Hershey, all the prime people, the major people who would go on to win Nobel Prizes for what they did, but every time those chromosomes split, they would go 'Oooh!' They would gasp. I didn't know if I should look at the film or the audience!"

"It was like a light going off in our heads," Maury Fox said. "Things were opening up. It was an excitement about the extent to which the inner life of the cell was being orchestrated." It was, Maury agreed, a very romantic time.

By 1950, there was one last basic discovery to be made. Nobody yet knew how genes directed the production of proteins, how DNA copied itself, and exactly how a mutation occurred. But since the structure of a molecule often opens the door to its function, most of the handful of scientists in the world who were starting molecular biology reckoned that deciphering the structure of the DNA molecule itself would be an important key to all three puzzles. Linus Pauling proposed a structure in the form of a helix with three chains, like rails, linking chemical units called bases, the railroad ties. Despite Pauling's stature as a Nobel Prize winner in chemistry, most thought his structure was a near miss. Francis Crick, and a young American named James Watson, decided to try for a better solution.

Both men worked at the Cavendish Laboratory in Cambridge, England, in a unit funded by the MRC. Watson, who had been a student of Hermann Muller and Salvador Luria at Indiana, was partly supported in his overseas studies by a fellowship from the National Foundation for Infantile Paralysis—the March of Dimes—the beginnings of what would become a trend of patient advocacy groups supplying money to researchers to pursue cutting-edge work.

At the time, Cambridge was still populated by many of the same names from Haldane's day. A Huxley, Hugh, helped Watson use an X-ray camera in his effort to visualize a crystal structure of DNA. Watson socialized with Avrion Mitchison, son of Naomi and Dick, nephew of Haldane. He spent house holidays at the Mitchisons' in Scotland. And among biologists, many of the same attitudes expressed by Haldane lived on. Watson described himself as "an escapee from Catholicism" and Crick was especially dismissive of religion. "The record of religious beliefs in explaining scientific phenomena has been so poor in the past that there is little reason to believe that the conventional religions will do much better in the future," Crick has written. Heaven doesn't exist. And if there's no heaven, if we disintegrate into nothingness, then it's better not to leave at all. "Not only do the beliefs of most popular religions contradict each other, but by scientific standards, they are based on evidence so flimsy that only an act of blind faith can make them acceptable. If the members of a church really believe in life after death, why do they not conduct sound experiments to establish it? They may not succeed but at least they could try."

There was also a self-consciousness among them that the work was of historic importance. As Watson wrote in *The Double Helix*, his cranky memoir of those days, some biologists had grown content and a little stymied by the complexity of genetics. "All that most of them wanted out of life was to set their students onto uninterpretable details of chromosome behavior or to give elegantly

phrased, fuzzy-minded speculations over the wireless on topics like
the role of the geneticist in this transitional age of changing values."
Watson and Crick, on the other hand, wanted to help start a new age
in which a scientist could actually do something with DNA rather
than talk about it.

Finally, in 1953, thirty years after Haldane's speech, ten years
after Schrodinger's *What Is Life?* and Oswald Avery's discovery of
genes-as-DNA, Watson and Crick, aided by London-based rivals
Maurice Wilkins and Rosalind Franklin, had the right structure. On
April 2, 1953, the journal *Nature* ran a short paper authored by Crick
and Watson entitled "A Structure for Deoxyribose Nucleic Acid."
Rather than Pauling's triple helix, Watson and Crick proposed a double helix with two twisted rails supporting the chemical bases adenine, guanine, cytosine, and thymine. Schrodinger had been correct.
Life was a process of physical chemistry akin to the chemistry
between the great comedy team of Abbot and Costello. Cytosine and
thymine, known by chemists as "pyrimidines," are the Bud Abbott of
molecules—short and chubby with one ring structure projecting out
from them. Adenine and guanine, the "purines," are more like Lou
Costello, elongated by having a second ring projection. Because of
this structural difference, the bases on one rail always paired up with
the same companion base on the other rail: adenine always paired
with thymine and cytosine always paired with guanine, the rings all
held together according to the laws of thermodynamics by bonds
made of hydrogen atoms. In DNA, you always knew "who was on
first." In a final sentence so laconic that it has become legendary,
Watson and Crick wrote that "it has not escaped our notice that the
specific pairing we have postulated immediately suggests a possible
copying mechanism for the genetic material."

In other words, structure did provide a key to function. Now, scientists could figure out how genes worked as a blueprint, directing
the machinery within a cell to make proteins that, in turn, talked to
other cells, and then tissues, and were ultimately responsible for
life—everything from height and eye color to health, memory, sex
drive, and longevity.

AMERICAN POPULAR CULTURE embraced the gene, and everything
else connected to science. In May 1948, a former science editor of

Life magazine, Gerard Piel, had brought out the first issue of the resurrected *Scientific American* magazine and delivered science to the
masses. It was a hit. And real science was long-jumping expectations,
enabling people to imagine science fiction coming true. Bill Haseltine, of course, could see the very tangible results of new technology
any time he walked outside his house in the form of contrails etched
across the sky by new airplanes and rockets.

He could also read about seemingly miraculous medical advances.
The production and use of penicillin boomed after the war, so treating
common infections like strep throat was as easy as a shot in the arm.
In 1951, when he was seven, polio was scaring the pants off parents all
across the country. Entire hospital wards were lined with iron lungs
wrapped around youngsters who could no longer breathe on their
own. But in 1952 Jonas Salk gave out the first dose of a polio vaccine
and mass inoculations followed. The fear eased. Science triumphed.
No wonder a kid growing up in what amounted to a science campus
would think that, given enough time, a little luck, and the money to
pursue the research, just about anything was possible.

By 1962, the year Bill Haseltine graduated from high school, you
could have sex whenever you wanted and not become pregnant,
thanks to the birth control pill. You could take a little Seconal to help
you sleep or just take the edge off after a bad day. Sean Connery was
playing James Bond in *Dr. No* and using all the best gadgets science
and technology could produce while having guiltless sex with gor-
geous women. America was becoming a shiny chrome and stainless
steel and plastic nation and better days were yet to come. No wonder
that when young Haseltine left China Lake to attend the University
of California, he was sure that science, and medicine in particular,
was the place to be.

Already he was so suffused with a dedication to science, and to a
vision of his own destiny, that every other passion in his life was a
secondary consideration. William Haseltine was not an average,
early 1960s University of California student who got decent grades,
listened to Pete Seeger, and tried to pick up sorority girls by singing
"Rock Island Line." He arrived in Berkeley carting his newly discovered intellectual precociousness, his self-consciousness as a winner,
and went straight to work. "I was an absolute Trojan," he says as we

sit in his living room in New York. "I was extremely intense about school, to say the least."

While other students grew goatees and hung out in coffeeshops trying to imitate the beatniks who crowded City Lights Bookstore across the bay in San Francisco, Haseltine woke up on time every morning, polished his shoes, put on a pair of slacks, a white shirt, a tie, a jacket. Florence was two years ahead of him at Berkeley and was doing well herself, and his father had taught there, so Haseltine had a legacy to consider. But he also had something to prove. During the first lecture of his first chemistry class a bow-tied professor stood in front of the class and instructed the freshmen to look to their left, then to their right. Only one of the three, he said, would still be in the university by the time the year was out. Haseltine decided he'd be the one left standing.

The professor was right. The classmate on Haseltine's left dropped out before the semester was over. The one on his right stayed longer, but eventually quit, too. Haseltine remained.

From his first days in Berkeley he was determined to prove that he could not only survive, but that he could be the best, even if it meant spending hours in the old chemistry labs. They were smelly, dark places that Haseltine loathed and the assigned work—a lot of repetitive experiments designed to demonstrate things like the way chemicals can change color, work that ranks of students had done for years—seemed pointless. The labs did have one very special virtue, however. George Pimentel, head of the department, required professors, not just grad students, to man them. Those professors formed a network of scouts who could spot budding science talent.

Pimentel was a Berkeley icon. He invented a key component to infrared spectroscopy, helped create the first chemical lasers, was a pioneer in space research. But in addition to his own work, Pimentel believed that it was his duty to throw an arm around the students and show special attention to those who displayed promise lest they stray from their science destinies. When the chemists who worked in the labs tagged a prospect, that student was asked to meet with Pimentel personally. Haseltine was one of the chosen. During their interview, Pimentel offered Haseltine a place in the department's intense summer program and Haseltine leaped at the chance.

That first summer students were required to spend each week studying the work of one of the top physicists or chemists in the world. At the end of the week, the students would meet the man himself. During those three months, Haseltine met Nobel Prize winners like Owen Chamberlain, the co-discoverer (with Emilio Segre) of the antiproton; Melvin Calvin, who figured out how plants use carbon dioxide; and Harold Urey, who had discovered heavy hydrogen in 1934. Haseltine learned all the facts, but there were more important lessons. Despite the accolades they had received and the years of experimental work on their resumes, none of the great men seemed jaded or tired. Though some of the breakthroughs had come before Haseltine was born, these scientists retained their fervor. Science, Haseltine learned, could be a way of life, almost a philosophy. And if you were good at it, fame could fall in your lap.

Every year thereafter, Haseltine spent his summers working in Pimentel's program. One summer he worked on a project to develop ways to find life on Mars. As part of the project, Haseltine partnered with a graduate student to work out methods for interpreting the spectrum of the Martian atmosphere. The paper produced from that collaboration was eventually published, providing Haseltine his first listing as an author in a science journal, a remarkable achievement for a college sophomore.

In Haseltine's final summer, Pimentel provided travel money so the students could go anywhere and study under whomever would take them in. Haseltine wound up at MIT, where Florence was now in graduate school and where their father and grandfather had studied. There he was assigned to spend the summer in the lab of Alexander Rich. Rich was twenty years older than Haseltine and was a rival as well as a confidant of some of the founders of molecular biology like Crick, Watson, and Walter Gilbert. He was a tie-wearing member of the "RNA Tie Club," the ultimate biology geek fraternity, a group of biologists and physicists who were attempting to work out how genetic information is translated into proteins in the cell and who wore ties painted with a model of an RNA molecule. Rich also had an impressive list of accomplishments. In 1956, while working at the NIH, he had discovered "hybridization," the ability to unzip the two strands of DNA and mate them to other unpaired strands according to the "who's on first rule" of base pairing to create new double-stranded helices. Such hybridization would later become one of biotechnology's foundation inventions. Then, in 1963, just two years before Haseltine entered Rich's lab, Rich led the team that discovered the polysome (or polyribosome), a cluster of ribosomes, the machinery inside the cell that produces proteins.

Cambridge, Massachusetts, was the white hot sun of molecular biology at the time, and it would remain so for another decade. There were important satellites, like the University of Wisconsin at Madison, Stanford, Michigan, Berkeley, University of California at San Francisco, Washington University in St. Louis, the original Cambridge in England, the Pasteur Institute in Paris. But the vortex of Harvard and MIT drew Haseltine with irresistible gravity as he came to believe that he wanted to, and could, compete in the arena with the likes of Rich. He still entertained thoughts of medical school—Rich had earned an M.D. as well as a Ph.D.—but pure, basic science was seductive. Pimentel's strategy had worked. Haseltine had been excited by science since junior high, but for three years Pimentel showed him greatness and now Haseltine's own ambition to become great had been confirmed.

The discovery of the double helix structure and work from the likes of Rich and the other giants of molecular biology were the end of the beginning of research into genes and DNA. But just as one of life's great secrets was revealed, science stepped back, at least publicly, from the grand visions of Haldane, Schrodinger, Wells. There were several reasons why. First, biology was looking more and more like peeling an onion. The unveiling of every new layer only revealed another layer. It was obvious now that any human redesign was going to require long years of work. Second, with the advent of molecular biology, scientists were finally able to work on the same minute, reproducible scales as the "harder" sciences of physics and chemistry. Biologists were no longer just observing what happened to cells in dishes or flies beamed with X rays. They were systematically taking apart the processes occurring inside cells. Haldane had skipped ahead and talked about the bigger picture—what could happen when you put those processes back together again in a new way—but now the scale of biology became much more intimate. Instead of Haldane's new societies, or Hammond's herds of refined animals, biology was now about molecules, and scientists reveled in focusing on one branch of one tree in a great forest. Finally, science was becoming much more institutionalized. Before the war, a man like Haldane could stand up and say any damn thing he wanted. He was beholden only to his college. But now governments and big institutions were funding science, and scientists who risked sounding like kooks also risked losing research money. With the "hardening" of biology, the culture of the tribe became more conservative, less speculative. Nobody talked about human immortality or enhancing human beings—at least not publicly.

Many in the culture at large, however, felt free to do all the speculating they wanted. They began looking to the future, to fill the imagination gap left by mainstream science. It was obvious to them that somehow, some way, science could fulfill the promise of Francis Bacon, and Haldane, and Schrodinger, and H. G. Wells, and if science was not prepared to take the initiative, they would.

4 ARISE, LAZARUS LONG!

WHEN HOWARD TURNEY was fifty-nine years old, he was hit with
the startling epiphany that his best years were behind him. He had
gotten fat, he had developed sporadic palsy, and, maybe worst of all,
he "was a washout in the bedroom." In the nick of time, Turney read
news of a small 1990 study in which elderly veterans were injected
with HGH. That single study, performed by an old friend of Florence
Haseltine's, showed that most of the men felt better. They lost fat,
their skin was more elastic, they gained muscle, their mood
improved. The investigator pointed out that there were significant
side effects to growth hormone and he did not recommend that people take it to achieve any of these improvements, but Turney saw a
way out of his physical deterioration. Turney lived in Houston,
Texas, so he could travel down to Mexico, where a fellow with the
right connections and money could obtain vials of growth hormone—which American doctors would prescribe only for dwarfism
and severe deficiencies. Turney began shooting up and the results
were, he claimed, miraculous. Just a few months after he began his
new regimen, a dental disaster created havoc in his mouth, necessi-
tating hours of oral surgery, multiple root canals, four pulled teeth.
But when he wound up back in the dentist's chair just eighteen days
later, his dentist was amazed. His mouth had accomplished two
months' worth of healing in two weeks' time. Turney decided right
then that he just had to share his secret with the rest of the world, at
least that part of the world that could afford to fly down to Mexico

and pay for pricey hormone injections. He established the El Dorado Rejuvenation and Longevity Institute and became a tireless salesman for growth hormone. Why, he said, not only did his physical ailments disappear, but he was now a sex god, back to his twenty-five-year-old stamina and drive with timber between his legs that was better than ever, and if you doubted his word, he had a girlfriend twenty years younger than he was to prove it.

While Turney's immediate inspiration for starting El Dorado was the published growth hormone study, there were already people around Houston interested in life extension and cryonics, and their original inspiration came not from some scientific study, but from a science fiction writer named Robert Heinlein. Turney, in fact, would eventually change his legal name to Lazarus Long, the immortal hero of Heinlein's "Future History" series of novels and novellas that also feature an age-reversing chain of medical facilities called the Howard Rejuvenation Clinics. Indeed, it was as Lazarus Long in January 2000, that Turney, then out of the rejuvenation business, would be slapped with an injunction by the Securities and Exchange Commission (SEC) for selling bogus shares in a new Caribbean country he claimed to have founded called New Utopia. His fine was waived because he was broke.

Heinlein, of course, was not the first person to think up a clinic for rejuvenation. The search for life-extending elixirs and treatments to restore youth is almost as old as civilization. The ancient Romans established precursors to European spas in places like Baden-Baden, Germany, and Bath, England. In 1931, Dr. Paul Niehans made the concept of rejuvention even more literal by starting Clinique La Prairie in Clarens, Switzerland, to exploit a technique he had developed called "cellular therapy." The technique was based on the work of the early developmental biologists who studied how animals grow from embryos into whole creatures. Those biologists had uncovered the ability of cells in an embryo to form the tissues of a body. So Niehans tried to harness this power by injecting cells from fetal sheep into human beings, the idea being that the fetal sheep cells would make the body's tissues young again. Niehans attracted an elite clientele—from minor nobility, to movie stars, to wealthy Americans, including a woman named Florence Mahoney who would become one of the most powerful health advocates in the United

States, and who, like the Houston life extensionists, would later influence the creation of the biotech religion. (Whether or not the La Prairie treatments worked, Mahoney did live a long life. She died on November 29, 2002, at age 103.) But while Heinlein did not invent the rejuvenation clinic, his philosophy found a waiting audience.

Heinlein, a native of Missouri, was a graduate of the U.S. Naval Academy. He left the navy in 1934, studied at the University of California at Los Angeles, and began writing and publishing stories in the pulps in 1939. Heinlein first introduced the world to Lazarus Long in a 1941 version of what would become a novella entitled *Methuselah's Children.* Long is Heinlein's fictional doppelganger, a rough-and-tumble man's man, a man who believes in science, rationality, plain speaking. Long is a womanizer. Long is libertarian. Best of all, Long is immortal.

Heinlein used the latest developments in biology, and speculations about biology, to carve out Lazarus Long's history. For example, Lazarus became immortal as a member of the "Howard families," a collection of eugenically derived, long-lived people descended from a few families whose matings were arranged by pairing up longevity champs, a system funded by a foundation. Eventually, the Howard families become feared by the short-lived, and so, led by Lazarus Long, they hijack a huge spaceship and travel to other worlds seeking a new home. They survive during the long intervals of space travel partly through a technique Heinlein calls "cold sleep," a state of suspended animation. Eventually, the Howard families return to earth. But they are surprised upon their reentry. The short-lived, deprived of vengeance upon the Howard families but inspired by the knowledge that the human life span could be longer than they thought, skip over eugenic matings and mount a worldwide research effort to develop rejuvenation biomedicine. They succeed through practices like tissue engineering and cloning, disciplines that existed only in imaginations when the story was written but which were all part of Heinlein's weave.

In Heinlein's universe, death, at least as long as one isn't bored with life, is the ultimate tragedy. "Can you understand the bitter, bitter jealousy of the ordinary man of—oh say fifty—who looks at one of your sort?" the leader of the short-lived people tells a Howard family member. "For not more than the last ten years of his fifty, he has amounted to something . . . And now, when he has reached his goal, what is his prize? His eyes are failing him, his bright young strength is gone, his heart and wind are 'not what they used to be.' He is not senile yet . . . but he feels the chill of the first frost. He knows what is in store for him. He knows—he knows!"

To do something about the terror of that knowing, however, requires bold action. But to Heinlein, most people are lemmings—stymied, fearful, herd-like—while the great men and women, the true visionaries, are gutsy pioneers. "That tiny fraction that hardly shows statistically is the *brain,*" Lazarus says in Heinlein's mammoth 1973 sequel, *Time Enough for Love.* "I recall a country that lost a key war by chasing out a mere half dozen geniuses. Most people *can't* think, most of the remainder *won't* think, the small fraction who do think mostly can't do it very well. The extremely tiny fraction who think regularly, accurately, creatively, and without self delusion—in the long run these are the only people who count and they are the very ones who migrate when it is physically possible to do so."

These people even have better sex uninhibited by pointless taboos. "Migrants tend to be both horny and easy about it," Long says in *Time Enough for Love.* "Superior intelligence always includes strong sexual drive." Long's drive is so powerful, he winds up having sex with his daughters, who are actually feminized clones of himself, and with his mother when he travels back through time. While that sounds ridiculous in the real world, in Heinlein's, it all makes perfect sense. It's all very reasonable.

Heinlein was not alone in predicting new technologies to extend life. In 1948, the pulp *Startling Stories* ran a bit of fiction by an obscure writer named Robert Ettinger entitled "The Penultimate Trump" about a man frozen during space travel and brought back to life. Ettinger had been inspired by a similar story, in which a man frozen in space is revived by a race of robots. But he thought the writer of that one missed the point. "If immortality is achievable through the administrations of advanced aliens repairing a frozen human corpse, then why should not everyone be frozen to await later rescue by our own people?"

While life extension was a sci-fi staple, it also penetrated mainstream publishing during the postwar years. In 1949, publisher Roger Straus of Farrar, Straus and Company (later Farrar, Straus and

Giroux) published a book by Gaylord Hauser entitled *Look Younger, Live Longer*. At the time, Hauser was a popular "expert" on nutrition and health practices, an adviser to celebrities. The advice in the book amounted to a collection of behavioral modifications and dietary suggestions, useless, mostly, for the purposes of looking younger or living longer. To his credit, Hauser did help popularize the consumption of yogurt as an antiaging health food, a theory later exploited by Dannon, which ran TV commercials in the 1970s showing smiling, toothless peasants in the mountains of the Soviet Republic of Georgia eating yogurt and laughing their way past 100. But the flimsy science in Hauser's book didn't matter. By 1950 it was a best-seller.

There was a hunger, especially in the most optimistic country in the world, the United States, to avoid death. People were willing to invest in it, to think about it, to investigate it. They wanted to be prettier, smarter, and up to date, too, the way the new kitchens and the streamlined locomotives were up to date. The future was where it was at and with the good news coming out of science and medicine, there did not seem to be any particular reason why we could not zoom into the future faster. While most people figured they'd just have to wait until science delivered the goods, a few, a motley and disjointed collection of individuals and groups, set out to call the future down from the sky, to make Francis Bacon's Bensalem a reality. By the time they were through, they morphed into the ultimate bioutopian salesforce.

MODERN TRANSHUMANISM IS A STATEMENT of disappointment. Transhumans regard our bodies as sadly inadequate, limited by our physiognomy, which restricts our brain power, our strength, and, worst of all, our life span. Transcendence will not be found in the murky afterlife of the usual religions, but in technological and biological improvement. "I'd like backup copies of myself," one told me at a transhumanist conference, "something more durable than a carbon-based system."

Nobody knows who is responsible for creating the idea of the "transhuman." Wells, maybe, though he never actually used the word. Or perhaps Bacon with his supermen in New Atlantis. A modern wish for an escape from the burden of nature may have its roots in spiritualism. Madam Blavatsky caused a sensation when she arrived in the United States in the 1870s and espoused her "Theosophy," a quasi-religion based on the belief in reincarnation and a conviction that the human mind could reach beyond the body. Spiritualist churches sprang up and people believed they could speak with the dead. (My grandfather was a member of such a church, often revealing the contents of sealed envelopes and directing people to lost objects.) Though science, "mesmerism," for example, was sometimes used to investigate spiritualism, and though Harry Houdini often tried to blunt the impact of spiritualism by revealing the tricks of mediums, the desire to believe that life was more powerful than the body's limitations was so strong that spiritualism became an established belief, the fringe movement of its day.

Within modern transhumanism, however, Julian Huxley, Aldous's brother and Haldane's old compadre, is thought to be the first to coin the word itself. Robert Ettinger was the first to call it into action.

In 1962, Ettinger published a manifesto he entitled *The Prospect of Immortality*. The book was picked up by Doubleday in 1964, and published in several languages and in several editions. It became the founding document of the cryonics movement.

Saul Kent, the man behind the Life Extension Foundation and the FDA Holocaust Museum, was twenty-five years old in 1964. Death was far away. But once he read Ettinger's book, he decided to become a cryonics activist. "I knew that my youth and health were short-lived; that I was programmed to grow old, suffer, and die; and that major scientific advances would have to occur to change all that. I saw cryonics as a dynamic, dramatic force to drive the pace of research forward." Once Kent looked into cryopreservation technologies, he was thrilled and amazed to find that "mainstream scientists" were already on the case. It became his mission to help these researchers by providing funds and, someday, research efforts of his own making. He imagined a day when immortalists would actually join with mainstream science. So united, Kent hoped, the two sides could create "the most powerful and far-reaching revolution in history" that would "lead to physical immortality and the opportunity to explore an incredibly vast universe of unimaginable riches."

But cryonics was a seat-of-the-pants effort run by optimists, a subculture populated by science fiction buffs with an ecstatic belief in the future and a gung-ho, Mickey Rooney, "let's-put-on-a-show" enthusiasm. In the real world, freezing dead people was a struggle. Mike Darwin, who skipped the typical teenage work experiences of frying burgers or lifeguarding so he could help Saul Kent's New York facility, Cryo-Span, freeze the recently dead, recalled some of the dark comedy of the early days by detailing the fates of individual patients in a 1992 posting to a cryonics Internet message board. "Herman Greenberg: a 51-year-old accountant who was placed into suspension by his daughter Gillian Cummings in May of 1970. Mr. Greenberg's suspension terminated when his wife ordered his removal from suspension after the death of their daughter in the Cryo-Span facility where she was living in her car. It should be noted that Gillian was living in the facility and working on an ice-cream truck to make enough money to keep the liquid nitrogen bills paid and her father and other patients at the facility frozen."

Unfortunately for Kent and cryonics, science ran the other way. There was indeed some interest from real scientists at first, but cryonicists seemed a little too zealous, too ready to shortcut the science in their eagerness to implement "cold sleep." Cryobiologists, who knew that the cryonicists were simply turning bodies into mush held together by ice, began to shy away. A few believers, like Kent, pushed for more real scientific research, but the movement was ill-equipped to perform it, did not have the money to fund others, and was too impatient to wait.

Yet the idea caught on. Cryonics was given publicity—most famously in Woody Allen's *Sleeper*—that was wildly outsized for the tiny number of people who were seriously interested in it. But the concept of somehow cheating death became a popular notion, something to be wished for just as Heinlein said it should be.

Feridouin M. "FM" Esfandiary was one of those captivated. Esfandiary was an Iranian intellectual who had, because of his father's diplomatic career, become a citizen of the world. He eventually moved to the United States, began a writing career, and then taught at New York's New School for Social Research. Esfandiary believed space travel and technologies like cryonics would not only help make him immortal, but would fundamentally change human beings. For the first time, he thought, man did not have to be an exclusively terrestrial being. Our biology was a product of the earth, but now we were beginning to leave the earth, a development that would have profound implications for our future evolution. Clock enough space time, or go live on some other planet, and eventually the human species would evolve into something else. Esfandiary thought this was a good thing, part of the human adventure. Earth, and human biology, were limiting. People died. They got sick. Even the smartest weren't smart enough.

About the time Esfandiary was beginning to preach a techno-utopian future, a Harvard University psychologist named Timothy Leary was finishing up an experiment called the Harvard Psychedelic Research Project whose members included, among other luminaries, Aldous Huxley. The point of the project was to limn the effects of psychedelic drugs, especially the new, man-made, technology-derived drugs like LSD. Drop acid in the right ways, under the right conditions, and maybe you could touch the face of God, whatever that meant, and for Leary, it came to mean fondling the strings of your own DNA. Leary believed in the "attempt to scientize myth and mythologize science," so with the right combination of drugs, you could transcend your limitations and turn your brain into God.

Leary was soon fired from Harvard.

But a new philosophy knitting together these various strands began to gel in the early 1970s. First, in 1972, Robert Ettinger released another book, *Man Into Superman*. It went far beyond the Heinlein notion of "cold sleep" to suggest that man should not only try to avoid death, but should modify himself with powers far beyond those of mortal men. By the following year, the year of Heinlein's *Time Enough for Love*, Esfandiary, Kent, and Leary were all in California, with Leary doing radio shows, Kent still involved with cryonics, and Esfandiary teaching futurism at UCLA, where his classes became known as "21st Century Gatherings," attracting a couple hundred people at a time. Many of them were simply curious, but some were converts to the faith of new human potential, a way of thinking that was at least partly influenced by Ettinger.

Leary developed a new slogan, recalled his friend and radio co-host Gabriel Wisdom. It was no longer "turn on, tune in, drop out," but SMILE, an acronym that meant Space Migration, I^2 (for doubling of human intelligence) and Life Extension. "Tim believed that thanks to new technologies the human race would evolve more rapidly," Wisdom said. "We were still just domesticated primates. We were at the larval stage."

Esfandiary published a trilogy—*Upwingers, Telespheres,* and *Optimism One*—to spread the word. Leary started quoting from *Upwingers* and Esfandiary started quoting from Leary. A few of Esfandiary's followers started calling themselves Upwingers and argued that science and technology would enable man to achieve a transcendence never before known. Esfandiary then set a precedent for those attracted to his philosophy by changing his name to FM 2030. The year 2030, when he would turn ninety-nine, would be a time of rapture—a time when immortality, a vast expansion of human knowledge, and the technology to create a techno-utopia ought to arrive. (FM 2030 may have been inspired to pick 2030 by F. E. Smith—Lord Birkenhead—who wrote *The World in* 2030 in 1930, a Haldane-esque book of predictions.)

In 1981, a young artist named Nancie Clark picked up an alternative L.A. newspaper and saw the handsome, chiseled face of FM 2030 staring back. She called him up and arranged a meeting. Soon an extended family was being created around FM 2030 and Nancie Clark—who became lovers—Timothy Leary, Pearson and Shaw, a few cryonicists. Clark mounted an art exhibit called TransArt and she and Leary cooperated on other projects. A local Los Angeles cable TV show spread the transhumanist word. There were certainly differences among this budding avant-garde, but they all shared a libertarian streak right out of Heinlein, casting themselves as the intrepid, in-the-know pioneers of the future.

By 1982, with the release of Pearson and Shaw's book on life extension, and their appearances on the Merv Griffin show and on local newscasts all over the country, the idea of taking a hand in one's own biological future attracted a mainstream audience. This was not a brand-new idea, of course. Gaylord Hauser was the Durk-n-Sandy of his day, and more recently Linus Pauling, the Nobel Prize winning chemist who had worked on the mystery of the structure of the DNA molecule, had been espousing something similar, by swallowing mega helpings of vitamin C. Pauling was eighty-one and still going strong when Pearson's book came out (Pauling would die at ninety-three just a few months before the 1994 Las Vegas A4M conference), so his own example appeared to support the idea that aging could be slowed through nutrition. But by the time Durk-n-Sandy were selling thousands of copies, there was a vitamin and dietary supplement industry ready to feed the demand and keep it going. Within ten years, that industry was racking up billions in sales.

The Durk-n-Sandy message went beyond extending life. They favored human enhancement and since nobody knew exactly how to do this with genes or some other medical technology, Pearson and Shaw concentrated on the Learyian message of "smart drugs." For example, hydergine, they said, could boost one's brain power and memory. Deprenyl could turbocharge your body's energy. It was not only possible to live longer, it was possible to live better.

The smart drugs message appealed to a few members of the San Francisco Bay area's fledging computer scene. In 1984, Ken Goffman, who had begun calling himself R. U. Sirius, worked up a small 'zine, *High Frontiers*, subtitling it "Psychedelics, Science, Human Potential, Irreverence, and Modern Art." *High Frontiers* appealed to the outlaw wannabes within the tech world, the "phone phreak" alumni who had tormented the phone companies by making free calls using illegal circuit boxes in the 1970s, who had since gone legit by working for up-and-coming computer and software companies in Silicon Valley. Right after the first issue was printed, Sirius attracted some money in the form of Alison Kennedy, whose family money, connections, and a one-time acquaintanceship with Aldous Huxley gave her the perfect set of credentials to help R. U. Sirius make a going, if irregular, concern out of *High Frontiers*. Kennedy took to calling herself Queen Mu.

The Bay Area breeze was blowing an intoxicating fog of silicon, drugs, biology, and a modern bohemianism, and a lot of people got high. Human immortality was coming. Human beings would soon transcend themselves by merging with their technology. A little known sci-fi writer named William Gibson started the cyberpunk movement by publishing a novel entitled *Neuromancer* in 1984, extrapolating from a Defense Department project called Arpanet that was designed to link computers, to create a future world in which people came equipped with computer jacks in their skulls and could become part of a massive worldwide brain. A hero of the budding transhumanist movement, the French mystic monk Pierre Teilhard de Chardin had imagined just such a transcendent global mind, and as Arpanet morphed into the Internet, the global brain took on an air of inevitability and the concept of the computer-augmented human seemed closer than ever.

As the semiofficial 'zine of record, *High Frontiers* was a perfect forum for FM 2030. "We interviewed him in 1985 in Santa Monica," R. U. Sirius told me. "He had a natural, spry optimism about humanity. He was not a scientist, but a popularizer. He picked up ideas and synthesized them. He was massively well read." The people around FM 2030 were all well read. "We ran into people during that trip whose libraries and collections of recent scientific papers filled up an entire house."

In 1989, FM 2030 wrote a new manifesto, *Are You Transhuman?* a kind of expansion of Ettinger's *Man Into Superman* based on developments in both the world of computers and biology. That same year, Sirius, Queen Mu, and a few others started a new magazine called *Mondo 2000*. Psychedelics, though still part of the social life, were deemphasized in favor of the idea that digital culture was going to do at least as much to promulgate a new civilization as ketamine, DMAE, or LSD. Bay Area alternative types, like former acid tripper Stewart Brand, who had founded a bible of the organic movement called *Whole Earth Review*, saw this confluence of technology and biology as the next cool thing. "There was really a sense of excitement, a fizz, a tingle in the air," Sirius said. "People were contacting each other all over the place, reaching out to Gibson, to people involved in computer software, gathering around our party scene in Berkeley. . . . It was a very fun time, like Paris in the 1920s. It felt that way. The sidewalk cafés of Berkeley." *Mondo 2000* eventually imploded, but by the time it did, it had inspired the creation of another magazine, *Wired*, which liberally appropriated *Mondo 2000*'s writers and its techno-positive attitude. By 1996, people like Brand and Kevin Kelly, a former *Mondo 2000* writer who became editor of *Wired*, would be considered mainstream prophets of the corporate future and *Wired* founder Louis Rossetto would be addressing the World Economic Forum in Davos, Switzerland, and the Cato Institute, the libertarian think tank. There, he'd tell the intellectuals that Pierre Teilhard de Chardin's prophecy of a transcendent human existence via a global brain was coming true. The future was here.

In the space of twenty years, the Internet, the computer, the fiberoptic cable, and the cell phone had all gone from exotic to required. The philosophers of the digital age walked triumphant upon the earth, and now that they had installed the digital machine as indispensable accoutrement, they turned their attention to the human machine and to what they believed was the coming merger between the organic and the silicon worlds. The thinkers from organizations with names like Global Business Network, Long Now Foundation, The Institute for the Future, The Foresight Institute, and Technology, Entertainment Design (TED) conference, and the transhumanist computer experts like Hans Moravec, a robotics innovator; Marvin Minsky, the director of the MIT artificial intelligence (AI) lab; Ray Kurzweil, a multi-award-winning inventor and author said it was time to move on to a great convergence of biology and machine, for what was man but the DNA-programmed, turbo-driven, CD-burning, fully multiplexed Pentium of biology?

The whole thing seemed so natural to them that Kurzweil even had a name for unaugmented people. He called them Mostly Original Substrate Humans (MOSH). MOSH was old school. You didn't want to be MOSH.

It did not take much imagination to come up with handfuls of possible augmentations that would create a new paradise. Sex could be an electronically mediated, virtual reality (VR) experience and you could be intimate with anybody anywhere over any distance. Genders would be a thing of the past, or at least easily modified, thanks to VR. Species, too. If you wanted to be an impala on the veld, just jack in and play. Why be old when you could be young in your head, thanks to the computer and the vast worldwide brain? Why not make yourself invincible with machine add-ons?

Whether the average person thought about the world in terms of MOSH or not MOSH, Kurzweil was onto something. People, and not just self-professed transhumans, were attracted to the idea that machines and new biology could alter their bodies.

THIS MÉLANGE OF LIFE EXTENSION, sci-fi, computers, cryonics, and biology coalesced in the person of Englishman Max O'Connor. When the Durk-n-Sandy book appeared in the United Kingdom, O'Connor, then an Oxford undergraduate studying philosophy and economics, not only read it, he decided to live by it. Max began lifting weights, taking supplements by the handful, and developing a code of personal betterment. He contacted Alcor, which encouraged him to bring the message of salvation through deep freeze to the British and Max promptly set up the first cryonics organization in Europe, Mizar, in 1986. In the late 1980s, Max, enamored with the individualism that seemed such a part of American culture, came to California to study for a Ph.D. in philosophy at the University of Southern California. Soon, Max had fallen in with the local cryonics/transhumanist crowd. He, too, changed his name. Max O'Connor became Max More to reflect his quest for more of everything, an ever expanding menu of human possibility.

When Max and Nancie Clark met at one of Leary's parties there was an immediate chemistry. "We liked each other's bodies," she recalled. "By that I mean the way we stood. I have a way of holding myself, Max holds himself well. Then how we moved across the room. We caught each other's eye. It was charismatic attraction, then secondarily pheromones . . . and then we were intellectually stimulated." They later married, Nancie changed her name to Natasha Vita-More ("more life"), and the two of them, both attractive, good-looking people, became the poster-couple for transhumanism.

In 1988, with friend and fellow graduate student T. O. Morrow (tomorrow, get it?) Max started a journal entitled *Extropy*, a name meant to signify the opposite of entropy. He and Morrow wrote an Extropian Declaration and Max issued a statement of Extropian Principles, a Heinleinian philosophy of betterment through technology, and the creation of a posthuman future in which human beings would cease to be entirely human at all. Instead, they would so effectively integrate with new technology that they would become a new species, an idea the Extropians would expand upon with encouragement from Moravec, Minsky, Kurzweil, and the speculations of nanotechnology theorists who argued that, soon, the world would be treated to an unprecedented accumulation of wealth and innovation, thanks to atomic-sized machines that would do everything from assembling food to playing Roto-Rooter in the plumbing of our bodies by reaming out plaques in our veins and germs in our blood.

Extropians became an offshoot of transhumanism, finding philosophical homes in *The Fountainhead* and *Atlas Shrugged*. Like Ayn

Rand, they were tired of being held back by nannying government and worrywart religionists. There is some debate within Extropianism and transhumanism about just how libertarian the movements are but they believe in the Heinleinian concept of glorifying brain power. They think that only a few people are smart enough and daring enough to accept the Extropian challenge and they will be the ones who are saved. The uninitiated, the retrograde *Volk* trapped by religious superstition and fear of the new, well, they will be left behind.

The Extropians were fully aware that some people found them and their agenda funny, sometimes even scary, but Extropians believed that, as the ultimate early adopters, they could help lead the enlightened world into the bright sunshine of the future. Big declarations about that future attracted some minor media interest to the Extropians, partly because they made for an easy story. Writing about them was like shooting fish in a barrel—the smart-alecky jokes practically wrote themselves. Sexy Natasha, whose art had a sensual bent, often displayed her form on web pages and in photographs. Both she and Max believed in the ongoing perfectibility of the body and, like many transhumanists, both were cryonicists. So journalists played up the looniness of their ideas—Ha Ha—or the element of narcissism in Max and Natasha's body building, supplement ingesting, antiaging routines. Fittingly based out of their Marina del Rey apartment in California, the Mecca of reinvention, they became the very image of a southern California couple living just one block from the sharpened edge of the continent, hanging on by their fingernails to the last bit of transcendent American dream. But in person, as long as you weren't talking about becoming posthuman, Max and Natasha came off as a smart, fun, slightly bohemian couple with unusual enthusiasms.

Other Extropians were a mixed bag. In general they had a tendency to overestimate their intellectual prowess. A few could be insufferable, regarding any challenge to their wildly optimistic claims for technology as the result of childlike ignorance. They also tended to worship slavishly actual scientists like Kurzweil, Minsky, and biologist Richard Dawkins, who coined the word *meme* in his book *The Selfish Gene*.

According to Dawkins, a meme (which rhymes with *gene*), is any concept that becomes part of the international cultural language, a social gene. Extropians were constantly trying to create new memes. They wanted the language of immortality, human transformation, and extreme self-determination to become ingrained values around the world, but the Extropians did not make their job any easier with that name and its space-cadet ring, and their insistence on saying things that made other people uncomfortable, Nietzschean-sounding pronouncements about metabrains and ultra humans and the ascendance of the intelligent. It sometimes sounded as if they were plotting a future world in which they could take revenge on every jock who ever made fun of the smart kids.

While all Extropians are transhumanists, not all transhumanists are Extropians. Many others, like Saul Kent, have no formal label. Some cryonicists deny being transhumanists. They simply want to live forever. Sirius has called himself "extro-punkian" to distinguish his beliefs from the more libertarian wing. But to some degree, all cryonicists, life extensionists, neo-eugenic philosophers, and nanotech speculators are transhumanists and there is a good deal of cross-pollination between the various groups. But whatever the category, transhumanism is an argument for Haldane's vision of mind over matter. Nature has done a decent but imperfect job so far, goes the thinking, so now it is time for man to direct his own evolution.

There was just one problem with the burgeoning transhumanist movement. It started as a semiaddled, occasionally drug-fueled, riff based on wild exaggeration and the fun you could have thinking it all up, a collective exercise in wishful thinking. "It was a work of imagination," Sirius said. This did not mean transhumanists were crazy or completely self-deluded. From the beginning, all the way back to Wells, for that matter, transhumanists knew their ideas sounded outlandishly goofy. Some, though certainly not all, had a sense of humor about themselves and their proclamations. Even when they hung out with FM 2030, and talked about these dreams, they understood that it was all science fiction, inspired by Heinlein, Gibson, and a new generation of sci-fi novelists like Verner Vinge. The very rapture they looked forward to, the day foretold by Wells and then by FM 2030, the day when AI and nanotech and biology would combine to create exponential and uncontrollable advances in intelligence and evolutionary possibility, had been named "The Singularity" in 1993 in a novel by Vinge.

But they did not suck their ideas out of thin air. They thought they could see into the future by taking all the puzzle pieces jigsawed out of new technologies and science journals and putting them together to form a semicoherent vision. Unfortunately, whenever they tried to make that future come true, to act out the fiction, they stubbed their toes. Cryonics was a good example.

The cryonics movement, shut out of mainstream science, failing to sign up many paying customers, and run by uncredentialed amateurs who often had no other significant occupation outside cryonics, devolved into seemingly endless rounds of inside bickering. There was even a brush with the law in 1987 when the Riverside County sheriff suspected that Kent preempted the Reaper by having his mother's head removed while she was still—if you wanted to get all technical about it—alive. Everybody was eventually exonerated, but there was an unseemly search for Dora Kent's head, which may or may not have found its way into the trunk of Saul Kent's car and gone for a road trip on Interstate 10.

The cryonics movement became obsessed with attracting a famous, and preferably rich, patient to give the idea credibility, much needed publicity, and a cash infusion. It came close in 1988 when Alcor suspended a television producer and writer named Richard "Dick" Jones, one of the creators of "Tootie," "Jo," "Blair," "Mrs. Garrett," and the rest of the girls' school gang of the sitcom "Facts of Life." But after a legal battle with heirs, much of Jones's estate slipped away from Alcor's grasp. No other big names seemed interested, with the possible exception of FM 2030, and he wasn't all that big a name outside futurist circles. When FM did not, as he had hoped, make it to 2030, and died at age sixty-nine on July 8, 2000, he was indeed suspended by Alcor as a "neuro" but he died in New York, far away from Alcor and thirty hours elapsed before the suspension process could begin. The brain begins suffering irreparable damage after ten oxygen-deprived minutes. It would not be until 2002, when Alcor placed baseball great Ted Williams into a dewar, that big-time fame and freezing would mix. Leary also had a date with cryonics but in 1996, at virtually the last moment, he decided against the liquid nitrogen dip. His ashes were eventually shot into space.

The other aggressive push to implement the transhuman future had come from the dietary supplement industry, inspired by Durk-n-

Sandy. But the more people learned about the life extension diet, the more it seemed clear that all the vitamin C in the world was never going to make you live longer. The industry boomed, and health food stores started packing in a chemistry lab full of supplements, but when mainstream science took a good hard look, many supplements were deemed worthless, further alienating scientists from their enthusiastic cheerleaders. The drugs that did have real effects, LSD, ketamine, the man-made psychedelics, became controlled substances or were banned outright as soon as the feds caught wind of their existence. And besides, when their users took time out to really think about the effects, they realized that dropping acid was mostly a trippy way to spend an afternoon rather than a way to turn your brain into God or worship at the altar of your personal double helix.

There was, however, something very real happening with DNA, something that turned FM 2030's head from outer space to human biology and something that the digerati around San Francisco found more interesting than drugs or even silicon. DNA itself was becoming a technology. "A different orientation happened," Sirius said. "We got more oriented toward the hands on rather than the visionary aspects of DNA. We got more inspired by the actual technology rather than just some abstract acid rumination."

The biologists working on that DNA technology usually knew very little about transhumanism, *Mondo* 2000, FM 2030, the whole counterculture-techno-bio-utopian shebang. They didn't know just how many people were out there rooting them on. But they did know a little something about sham science and outrageous claims and they stayed as far away from it as they could, especially when it came to notions like life extension and human enhancement. The conservatism that had seeped in after World War II had become a dogma within biology. As a collection of mostly nonscientists—the computer gurus being a notable exception—transhumanists could afford to say whatever they wanted about the human biological future. But in biology, one did not extrapolate. You published only what you knew, you made no claim beyond what the experiment proved, because if you deviated too much from the accepted wisdom, you could kiss that grant money goodbye. Most of biology rewarded small steps, not big leaps, especially when it came to the potential for a human treatment.

But one realm that bubbled up from biology followed a different set of rules. It was a place where conservatism was deadly and where the big statement was high-octane fuel: biotechnology. Biology for profit. It's what turned FM 2030's head and made the predictions of Heinlein seem imminent. Biotech scientists and entrepreneurs were not just talking, stoked on dreamy sci-fi, a fist full of supplements, and a new Apple Macintosh. They were walking the halls of the biology buildings at Harvard, MIT, Cal, Pasteur, and Cambridge and they were wearing Brooks Brothers' suits and signing checks for millions. Biology had changed since Haldane's day, and it was more conservative about what it said and whom it embraced. But ever since *Daedalus*, it had been dissecting how life worked. And now a few, like William Haseltine, were saying that it was time to put it all back together again.

5 THE IMMORTAL MR. STEINBERG

ALTHOUGH THE TRANSHUMANISTS and the life extensionists had germinated an ideology, they still weren't much more than the sci-fi equivalent of a rotisserie baseball league. At the time, they had neither the organization, the science wherewithal, nor the financial clout to effect the transition to any posthuman future. The manifestation of their dreams would come from real science, bravado, and, most of all, money. And once those forces aligned, it would no longer just be transhumanists imagining a new future, but average folks who thought FM 2030 was a radio station.

The first of the triad of necessary ingredients for building the transhuman future was already well under construction by the time Bill Haseltine got to Harvard. The social upheaval of the mid-1960s was mirrored by a scientific revolution. The halls of university biology buildings were filling with brilliant people with new ideas and huge egos, some of whom, when they laid their heads on their pillows at night and thought about the big picture—as Haseltine always did—understood just how profound their work could be. They almost never publicly said that they might one day effect a transformation of man, either because they thought such a transformation was so far away as to be meaningless or because they had no desire to stir controversy. But by manipulating the most basic bits of life, they were harnessing forces to alter the course of evolution, just as Schrodinger and Haldane had predicted.

Haseltine himself made a conscious decision to take part in the revolution—both inside and outside the biology labs. He came to Harvard still thinking he wanted to be a doctor. His application to medical school had been accepted. But after his immersion in basic science with Pimentel and his experience in Rich's lab, and also because of some opposition from his father, who despised the idea of his son becoming a doctor, Haseltine waffled after he arrived in Cambridge. Then, a series of interviews at the medical school convinced him to give up the dream for good.

At the time of his interviews, a scientist and M.D. named David Hubel was working with monkeys in an effort to explain how vision operated in the brain, work that would eventually earn Hubel the 1981 Nobel Prize for medicine. Hubel himself had become disaffected from the practice of medicine during his hospital training and on the day Haseltine met with him, Hubel said, "You don't want to put your hands on sick people. You can do more good for medicine in the sciences. Understand as much science as you can and apply it." Haseltine never looked back. He seemed to know intuitively that his impact on the future would be determined by his ability to help alleviate, maybe even end, human suffering. Many people try to do their bit, try to ease the pain of this person or that one as a doctor does. But Haseltine wanted to do more, and do it bigger, and even when his thinking about revolutionizing medicine was vague, a thesis for his life without the outline, he knew he needed to be ambitious. Now he decided that one could be a car mechanic and provide a valuable service. Or one could make a better car.

Haseltine was not alone in thinking grand thoughts. Harvard and MIT just down the Charles River—where Florence was also in graduate school—were grand places. Big science was well established and Cambridge was one of the most powerful research centers in the world. And it was a heady time not unlike Haldane's 1920s when, following a period of conservativism, society seemed to be opening up to new ways of thinking.

Haseltine, for example, soaked up much more than science at Harvard. It was not unusual for him to dash from a John Galbraith lecture on the new industrial state into a chemistry hall, and then back to a talk by Stanley Hoffman on war. One day he sat in on an international relations lecturer he found so fascinating he insisted his friends come to listen, too. The speaker was Henry Kissinger.

He began attending Kissinger's talks regularly. Then winter set in. One day, discouraged by the cold outside, he stayed in his seat after one of Kissinger's lectures and listened to the next professor, Walter Gilbert, speak on biophysics. "I was suddenly more interested in what Wally was saying than in what Kissinger was saying and at the end of his talk, I went up to Wally and said, 'Wally, if I want to work in your lab, can I do it?' And Wally said, 'Well, you'll have to ask Jim.'"

"Jim" was James Watson, co-discoverer of the double helix. Watson had come to Harvard in 1956 as a young professor. He had also written the foundation textbook, *The Molecular Biology of the Gene*, in 1965, the year before Haseltine had arrived at Harvard, and now Haseltine was working under his supervision. Gilbert, the son of a Harvard economist, was a physicist, but in an example of the fulfillment of Schrodinger's plea to unite the sciences, he joined a Watson team in 1960 and had helped to isolate messenger RNA, the intermediate molecule between the genes and the protein-making ribosomes. He then switched completely over to biophysics and made a number of key discoveries, including (with Alan Maxam) a way to determine the sequence of the four bases in DNA (adenine, cytosine, guanine, thymine—ACGT), for which he shared the 1980 Nobel Prize in chemistry with England's Frederick Sanger, who had independently worked out his own sequencing technology.

In an example of the "hardening" of biology into a new kind of science, biophysics under Gilbert was designed to take physical science students, who may never have had any training in biology, and inject them into biological lab work. The idea intrigued Haseltine, who was eager to work in an area that would have some application to human medicine. Once Watson gave the okay, Haseltine placed himself firmly on track to earn a Ph.D. in biophysics just like Florence was doing at MIT.

The work of the molecular biologists of this period is now considered basic. It's taught to undergraduates and bright high school students. But at the time these men were revolutionaries. Watson, Mark Ptashne, Matthew Meselson, Gilbert, Crick, Joshua Lederberg, Jacques Monod and Francois Jacob in Paris, Sydney Brenner in Cambridge, England, all of them were an avant-garde, and they knew it. They thought biology had muddled along for ages, but now they were seizing control, determined to pick up the various strands of research that had been laid down over the previous decades and unite them to create the biological equivalent of physics' "grand unified theory."

The times were so exciting, even Cambridge social life revolved around biology talk. Every year on December 30, Sally Fox's birthday, a crowd of biologists would gather in the Foxes' small Victorian house. Florence, who was Maury's graduate student until 1972, would come. Her fellow grad students Ray White and David Botstein, both of whom would later become science and biotechnology superstars, came too, and so did Meselson, and Gilbert, and a young hotshot named David Baltimore. "They were big, big parties," Sally recalled. "People would spill down the staircase and into the kitchen and they would dance! I don't know how they did it, but they would." The Foxes frequently held open houses on Friday nights, too, and Baltimore threw parties of his own. Always, the air was thick with chatter, inside lab intrigue, and plans for future experiments. The members of the tribe would get drunk, talk theory, then go back to the lab and start working again.

There was a self-awareness that the big payday after all the work of the Oswald Averys, the Hermann Mullers, the Thomas Hunt Morgans, the Watsons and Cricks, all the physicists-turned-biologists standing in their moldy sneakers at Cold Spring Harbor in the 1940s and 1950s was around the corner. In 1969, Harvard's Jonathon Beckwith became the first to isolate a single gene. In 1971, a young postdoc named Inder Verma, while working in Baltimore's lab, used RNA to create a copy of DNA, a so-called cDNA, so that now biology could build entire libraries of genes. Haseltine entered Harvard just thirteen years after Crick and Watson published the structure of DNA, and now, for the first time in history, university researchers were fiddling with the most basic elements of life.

True, the work was almost solely confined to tiny bacteriophages and bacteria, but Jacob argued that "what is true for *E. coli* is true for the elephant." Life operated according to a unified set of rules. Jacob's dictum later turned out to be an exaggeration, but at the time biologists thought they were peering directly into the mechanism of all life and manipulating it. As if to confirm how fast the science was moving, just three years after Beckwith's isolation of a gene, two scientists from the Salk Institute for Biological Studies in San Diego, California, Theodore Friedmann and Richard Roblin, wrote an article for *Science* entitled "Gene Therapy for Human Disease?" in which they proposed to apply this knowledge to people by fixing or replacing defective bits of the human genetic code, defects that might be responsible for as many as 1,500 human maladies.

So much hope was placed on the march of biology that, in 1971, Richard Nixon declared a "War on Cancer," the seemingly insoluble human plague. Nixon was advised on this policy by a former academic named James Cavanaugh, an expert on health care administration and economics from the University of Iowa who had since joined Nixon's staff. Cavanaugh, who would later play an important role in Haseltine's life, was partly responsible for an enormous boost in the budget of the NIH and for advocating the training of new biologists. Soon, a small army of molecular biologists was being trained at universities all over the country. (Fifteen years later the ultimate molecular biology program, the Human Genome Project, would be justified on the grounds that it would assist the ongoing cancer war.) The futurists, too, insatiable consumers of science journals, were well aware of the state of the science. The same year gene therapy was proposed, Robert Ettinger wrote *Man Into Superman*.

Science was mirroring society. The idea that society could be arranged, improved to increase the happiness of man, rather than allowed to flow willy-nilly was an increasingly popular notion among the new legion of biologists, including Haseltine. Back at Berkeley, in 1964, Haseltine had briefly participated in the free speech movement, but at Harvard he made politics a second career. During a student-led anti–Vietnam War strike on March 4, 1969, George Wald, a Harvard biologist and 1967 Nobel Prize winner, gave a speech at MIT entitled "A Generation in Search of a Future." That future, he said, was not set. It could be altered, formed to be whatever the young people wanted it to be. Haseltine was among the many who were inspired.

As Haldane had pointed out in 1923, science was a radical endeavor by its very nature. The whole point of science was to challenge existing thought and either confirm or upend established truth, to never accept the status quo. But that requires a certain amount of hubris. Scientists in Cambridge not only possessed an adequate supply, success depended on it. Aggressiveness, arrogance, chest thumping were as important to a student's quiver of gear as books and microscopes and glassware.

Harvard historian Donald Fleming captured the feeling in a 1969 essay for the *Atlantic Monthly*, "On Living in a Biological Revolution." There was a new "underlying revolutionary temper" in biology, Fleming wrote, citing Lederberg, Crick, and Edward Tatum as generals, along with Robert Edwards, who was then conducting his early IVF experiments.

> Biology has always been a rather loose confederation of naturalists and experimentalists, overlapping in both categories with medical researchers. Today even the pretense that these men somehow constitute a community has been frayed to the breaking point. At Harvard, for example, the revolutionaries have virtually seceded from the old Biology Department and formed a new department of their own, Biochemistry and Molecular Biology. The younger molecular biologists hardly bother to conceal their contempt for the naturalists, whom they see as old fogies obsequiously attentive to the world as it is rather than bent upon turning it upside down.

A certain amount of ego was to be expected. "After all," Fleming wrote, "modesty is not the salient quality to be looked for in the new breed of biologists. If the world will only listen, they know how to put us on the high road to salvation."

Indeed, recalled biologist Jerard Hurwitz of founding fathers like Delbruck, Luria, and Watson, "they called themselves the church, and they were the popes."

Like the early Christians, these revolutionary biologists were filled with hope that their work would allow human beings to stride into a transcendent future. "The future may or may not belong to those who believe in it, but cannot belong to those who don't," Fleming wrote. The future according to those around Bill Haseltine belonged to molecular biology, which had the power to salve human desperation over disease and death. "The new biologists are the only group among our contemporaries with a reasoned hopefulness about the long future," Fleming declared.

Fleming was describing the conflict between the new reductionism—man as a clockwork—and the old naturalism, which saw man and all biology as part of an organic whole that could be observed, but rarely, if ever, controlled. The reductionists were winning the day precisely because they did have hope for the future. They hoped that what they were learning could help perfect man, or at least fix or prevent much of what ailed us. They were activists. But because they wanted to apply what they learned, they eventually wanted to integrate reductionism and naturalism, to finish tearing biology apart so they could put it back together, or, as Haseltine described Watson's program, to build bridges from genes to cells to tissues. Once those bridges were built, there would be practically no end to what you could do with human biology.

Those bridges were already under construction and had been for a long time in the parallel science world of developmental biology, the realm of Fleming's "naturalists." When T. H. Morgan sliced and diced planaria, and then watched the pieces grow into whole new animals, he was watching the power of cells to stop being what they had been, like muscle, say, and to direct themselves to do something else for the purpose of regeneration, like build nerves. Experiments during the 1920s showed that cells in a newt embryo "hear" signals that tell them what they are to grow up to be, skin or liver or brain. In 1962, University of Cambridge scientist John Gurdon took a cell from a tadpole's intestinal tract, injected it into a frog egg, and got a clone. The original cell had heard genetic signals that told it to stop being just a cell that worked in the intestinal tract and to make a whole new body. Later experiments with human cells showed that if you fused two human cells from different tissues together their genetic signaling would change and the cells could take on new properties.

Science like this would eventually motivate Haseltine to argue for a whole new field of medicine that would deliver immortality, but at the time, these experiments fell like snowflakes on the warm highway being paved by genes and phages and bacteria. It took the work of another developmental biologist, Robert Edwards, and an obstetrician, Patrick Steptoe, to make the reductionists look up from their *E. coli*.

Using techniques derived from the pioneering work of people like John Hammond, the man who brought artificial insemination to farm animals, Edwards and his co-workers managed to join human eggs and sperm in a dish. In succeeding experiments, they got the resulting embryos to grow. Watson, among others, took notice and declared that if the experiments ever managed to produce a live baby, the world could very well be on its way to making cloned human beings. He wrote of this possibility in 1971, the same year that Heinlein was writing *Time Enough for Love* and making human cloning and tissue regeneration part of the menu for creating immortality at the fictional Howard Rejuvenation Clinics. For the first time since the early part of the century, science seriously began to consider a mechanism for man to direct his own evolution.

THE BRAVADO NEEDED to make the big statements and to roil the waters of science and society was bred into Haseltine. Academic life in China Lake had been competitive. The University of California was a gauntlet that claimed as many victims as Haseltine's first chemistry professor said it would. But graduate school with the likes of Watson, Gilbert, and the coterie of other giants populating Cambridge was a blood sport. Students like Haseltine were taught to challenge every scientific paper aggressively while vigorously defending their own with a cocksure attitude and an unyielding faith in their own brilliance. "Usually we would start by saying that the guy who wrote the paper was a fool, we're a lot smarter, let's reinterpret, let's see what this benighted fool really means," Haseltine tells me, laughing. "That is how they thought and that is how they taught and how they acted. It was a very arrogant way to do it, but every one of them was extremely arrogant. Still, it was very productive. It was a wonderful way to get educated."

Students attended weekly seminars in which some would present and then defend their work. "Not only would Watson and Gilbert be there, but Matthew Meselson would be there and Mark Ptashne, so we would have four of the greatest minds in our field sitting there criticizing our work as graduate students . . . you learned to be pretty tough facing that front row. After facing them, nobody else could ever intimidate you."

Because so much of molecular biology was terra nova, great prestige could accrue to those who made breakthroughs—there was therefore enormous incentive to make big, bold statements, and then to guard one's territory jealously by proving naysayers wrong. Before Haseltine was through with Cambridge, he would become an avatar for the system, an enfant terrible willing to take on anybody, anytime, anywhere and to insist that what everybody thought could not possibly be true, *was* true. For example, in 1970, an English scientist named Andrew Travers had discovered a substance critical for the life processes of bacteria. "That was an enormously important discovery," Haseltine recalled. "It was featured in *Nature* and Andy became an immediate international celebrity."

Walter Gilbert gave Haseltine the task of purifying the factor. But Haseltine couldn't do it. So he tried to re-create the process the factor was supposed to initiate and was shocked to find that the factor was not required after all. "So I went to Gilbert and said, 'Look at this.' He said, 'Well, you should believe your results. Go back and get DNA and the same reagents Travers used; they're still in the fridge.' I did and got the same results." So, in early 1971 Gilbert sent Haseltine to Cambridge, England, to work directly with Travers.

When Haseltine tells me this story, he begins to squirm toward the edge of his couch. He uncrosses his legs and leans forward as if girding for an epic. I can see his blood rising, his eyes widening like a small town retiree's recalling the time he caught the game-winning pass in his high school's homecoming game. In the hours we had spent together, and in the hours we will spend together in coming months, he has never been so obviously excited.

"It was a very tense two weeks in Cambridge. I worked under Travers's supervision and conducted these experiments, and he would take my results in little tubes and analyze them and not give me the answer. I then flew back and had to go directly to a scientific conference. Travers had given me a sealed envelope with the answer inside it to take with me to the conference and to give to the conference chairman. Well, I got up and gave my talk and said, 'I tried to replicate these experiments and here is my data that supports the fact you [don't need the factor].' Three leading molecular biologists besides Watson and Gilbert were sitting there: Ekkehard Bautz, Charles Weissmann, and Jerard Hurwitz. Weissmann got up and said, 'I did those experiments and I got Andy Travers's results.' Bautz stood up and said, 'I did those experiments and I got the answer Andy Travers did.' Hurwitz said, 'I did those experiments *with my own hands* and I got the answer Andy Travers did.' So I said, 'Look, I did not make up this data. I told you how I did the experiments, and

I did them with the same reagents Travers used.' And then the orga-
nizer said, 'I would now like to read the letter from Andrew Travers.'
And in the letter Travers said, 'I'd like to apologize because Haseltine
came to my lab and did the experiments and we did not get the
results I thought we would get and I am going to have to try to sort
this out.' That was pretty dramatic!"

Haseltine leans back in the couch, surrounded by every traditional measure of success: a fancy weekend apartment in New York
in a high-profile hotel across the street from Central Park and
around the corner from Barneys. There is good art on the wall, and a
maid serving coffee. He wears khaki pants and loafers with no socks
the way wealthy people do on the weekends when they are lounging
about the house. But right now he doesn't care about any of that.
Right now, he is leaning back, smiling, savoring an instant when
everybody who thought they were smarter than him believed he was
wrong and had to eat their doubts. It is a quintessential Haseltine
moment.

But Haseltine irritated people that day. "It turns out that in science nobody likes anybody who disproves anything. It is much better
to prove something than disprove it, but that way it's hard to learn
anything."

In addition to earning a reputation as another arrogant reductionist, Haseltine was also getting a name as a rogue who was
extremely competitive. His thesis project, on a system of genetic
control in bacteria, became a competition between himself and
other researchers, a competition he won when the work was turned
into a 1972 paper for *Nature* and he was listed as the main author, his
first step into the science big time.

As soon as a biologist earns a Ph.D., it becomes vital to find a
postdoctoral research position, an apprenticeship during which,
under the tutelage of an elder, a researcher can make a name for
himself the way Travers had done. Getting into the right lab can
make or break a biologist's future. Haseltine shopped around for the
most high-powered one he could find that might lead to work with
some direct application to human medicine, so he looked for labs
conducting experiments with animal, as opposed to bacterial, viruses
because animals were more complex models. He applied to three:
Walter Eckhart's at the Salk Institute in San Diego, Paul Berg's at

Stanford, and David Baltimore's at MIT. Berg turned Haseltine down. Eckhart and Baltimore accepted him. Haseltine chose to stay in Cambridge partly because he and Baltimore shared antiwar political beliefs, but mainly because of his admiration for Baltimore's science. Several years earlier, Baltimore had helped discover an enzyme that allowed a virus to turn its RNA into DNA, an achievement for which he would win the Nobel Prize in 1975 while Haseltine was working in his lab.

He had worked under Watson, Gilbert, and now Baltimore, all of whom were, or would be, knighted by the Nobel, and now Haseltine himself wanted to make it to the top. Haseltine continued to play the science game with a take-no-prisoners style, alienating some of his lab co-workers and even some who worked in nearby labs.

He also continued to challenge anybody who challenged him. When he left Baltimore to start his own operation at Harvard's Dana-Farber Cancer Center, he collaborated with an MIT scientist named Dennis Kleid. Their research, published in 1976, showed how the genes of certain viruses were copied. Haseltine and Kleid argued that the duplication occurred in a particular order. Some did not believe it. Haseltine, who was doing the work with funds from the NIH, took on the NIH grant reviewer who visited the lab. The reviewer, a Berkeley scientist named Peter Duesberg, was famous for isolating the first cancer-causing gene. But Haseltine didn't care who he was. When Duesberg told Haseltine that the work was all wrong, Haseltine contradicted him. The two began to wrangle. "Everybody who was on the program project grant turned white. I was fighting with one of our reviewers! You just don't fight with a reviewer. But he was wrong! And so we had a big fight." Haseltine was eventually proven right. Duesberg went on to become the guru for those who would later argue that HIV does not cause AIDs.

Haseltine reveled in being the iconoclast. When his Dana-Farber lab team worked up a new theory of DNA damage, he presented the data at a scientific meeting and was met with stony silence from the audience because he had just told them that the theories they had developed over decades were incorrect. "Nobody clapped. They always clap. But nobody clapped. Then I got a hostile question. Then they clapped." But Haseltine's theory turned out to be right "and that was a lot of fun." When his lab focused its attention on

DNA repair, "it turned out that twenty-five years of textbook chemistry were wrong and wrong in every detail. . . . We showed how it actually happened." Whether or not Haseltine embellished the significance of this work is open to dispute, but it is certain that he racked up enemies. "If you upset somebody's apple cart they are unhappy. People whose oxen are gored will never forgive you."

Twenty years after events in Cambridge, biologists still talk, and argue, about Haseltine and his seeming inability to play well with others. His fame for his science was growing within the tribe, but so was his fame for bad behavior. He was increasingly regarded as being wildly competitive, of failing to credit the earlier work of others or even the contributions of those with whom he collaborated. (In a conversation about her own impressive record, Florence, the director of the Center for Population Research at the NIH, referred to an important women's health initiative and said, "What the hell, I'm a Haseltine. I'll take credit for it.") During one such episode Haseltine even defied Watson, who had left Harvard to take over Cold Spring Harbor but who was considered a powerful eminence. Watson asked him to make way for another researcher who was pursuing an important question about viruses, but Haseltine refused, turning his new lab's attention to the same question and then, say his detractors, taking credit for finding the answer even though the other scientist had published first. One scientist left the study of viruses altogether and became a developmental biologist after an especially unpleasant encounter with Haseltine.

Years later, when Haseltine was up for Harvard tenure, such behavior haunted him. In fact, the committee standing in judgment turned him down for tenure—a status that not only lends prestige but also virtual lifetime job security to academics—until Baruj Benacerraf, then head of Dana-Farber, intervened on his behalf. "In my case they finally convened a special panel of top scientists and doctors to review all the cases and found no merit, just a bunch of nonsense," Haseltine told me. "The conclusion was 'He is brash, he is bold, but is he any worse than other people? . . . [He's] probably better than some of the people he was trained by. Was Watson awful? Yes. Was Gilbert awful? Yes. Was Ptashne a pain in the butt? Yes. So what's new?"

Others, including some familiar with the proceedings, strongly disagreed, but there is no consensus on whether Haseltine was just a slightly exaggerated product of the culture as he claimed or uniquely guilty of sin. There is no doubt, however, that he was not the only competitive biologist. Some in his subspecialty of virology, the study of viruses, had honed competition to a fine art. One, an NIH scientist named Robert Gallo, worked on a type of virus called a retrovirus, just as Haseltine was doing at Harvard. Gallo thought that, contrary to mainstream opinion, retroviruses caused human disease. Haseltine thought so too, and in the course of his research, he began to meet a few others who were inclined to agree. Eventually, he was introduced to Gallo and, through Gallo, he encountered a science scrum within the NIH that made any rivalries he had encountered before seem like a girls' powder puff softball league.

"Talk about a nest of vipers!" Haseltine laughed. "They were all in each other's pants fighting like crazy, trying to screw each other. . . . What a scene that was!"

Donald Francis, then a scientist with the Centers for Disease Control and Prevention, recalled the atmosphere. "They seemed to hate each other. They are set up on the academic, kill 'em, fight 'em mode of operation, not on the collaborative mode."

But Haseltine thrived in this group, partly because he was an outsider from Harvard, partly because he also had an instinct for the jugular, and partly because they all shared a uniting belief that looking at natural disease in people was better than looking at diseases created in lab animals.

The competitiveness was natural for the same reason it was so prevalent among the molecular biologists. They were digging a new mine. If retroviruses did cause human disease, whoever had staked the first claim would be entitled to a fat gold vein. But being first was vital. If you were first to tap the vein, all important grant money would flow and your small lab would suddenly become a big one. Prizes and prestige would follow.

And now there was an added incentive to be aggressive in staking a claim: money. Biotechnology was being birthed and it allowed scientists to complete the triad necessary to fulfill the transhumanists' dream. Biotechnology allowed real science to turn bravado into cash.

IN 1971, FOUR MEN in northern California, a biochemist named Ronald Cape, a doctor named Peter Farley, another physicist-turned-biologist named Donald Glaser, and a venture capitalist named Moshe Alafi, decided that a new technology invented by Glaser might make some money. Glaser had found a way to select the most desirable bacteria from a large batch. This was an important innovation because products like antibiotics and certain vitamins were made by fermenting bacteria in big tanks and then collecting their output, a miniature version of cows on a dairy farm. Glaser's gear could help choose the best cows. Once it had, a biologist could then breed to make even better producers.

The four started a company they christened Cetus, derived from the Greek word for whale. This was the first time that basic scientists, scientists whose research aims were not directly related to making a product, were recruited by a venture capitalist into a start-up company to turn that basic science into a business, the model all early biotechnology would follow.

Many scientists who heard of the Cetus idea resisted it. When the company went to find more biologists to staff the research and production facility, some demurred, insisting that serving mammon was incompatible with serving science. Haseltine was utterly uninterested. He knew Glaser from his days at Berkeley and, as a student, he had seen prototypes of the machines that would form the basis of Cetus's creation. Though he vaguely understood that Cetus had been founded, he was far too absorbed in his Ph.D. thesis to pay much attention and in any case, in the 1970s, Haseltine was a confirmed "radical." "It was very unlikely that I would have considered anything like that. I was deeply opposed to anything that was close to capitalistic, to put it mildly. I was interested in medicine, but not interested in 'the system,' as we called it."

Venture capital, however, was very interested. Most venture firms had been started after World War II and were dominated by family fortunes. Alafi was one of the first to recognize that a new kind of business at an early stage of development but with an interesting technology could be supported for enough time to convince others to place their bets, too, either by attracting follow-on investors or by mounting a public offering.

At the time, big industry was trying to diversify and Alafi knew that, with big industry budgets and a creeping decline of industrial research and development, they could afford, and needed to, take gambles on such upstarts. He convinced companies like Standard Oil of California and a distiller, which could benefit from technology that created more productive bacteria and yeast, to invest in Cetus. He even got a Canadian mining company, International Nickel (Inco), and a Bay-Area venture firm called Kleiner-Perkins to invest.

At just the time Cetus was getting off the ground, the Stanford laboratory of Paul Berg, the man who had turned Haseltine down for a postdoc lab appointment, was conducting a series of experiments using the natural base pairing mechanism of DNA to weld two disparate pieces of DNA together. The paper that resulted from this work, published in the *Proceedings of the National Academy of Sciences* in October 1971, was entitled "Biochemical method for inserting new genetic information into DNA of Simian Virus 40: circular SV40 DNA molecules containing lambda phage genes and the galactose operon of *Escherichia coli*." Hidden in that title was the fact that Berg and his co-workers had invented gene splicing— genetic engineering—or, as it's known in biology, recombinant DNA. Berg won the 1980 Nobel Prize in chemistry for it, an honor he shared with Gilbert and Sanger, who won for their DNA sequencing work.

At first, gene splicing was almost prohibitively difficult to do, but the following year, Berg's lab figured out a simpler method and that same year, a UCSF scientist named Stanley Cohen and a Stanford biologist named Herbert Boyer began to collaborate in an effort to combine the accumulated results from experiments in their labs and Berg's lab to create a recombinant DNA product that would be predictable, would have just the DNA they wanted, would incorporate that DNA into a bacteria, and would make that bacteria produce the predicted protein from that DNA. Over the next two years, they accomplished all those goals. Haldane was just a few years off in his prediction that science would make designer life.

In theory it was now possible to add a gene to bacteria. The gene would be read by messenger RNA. The messenger RNA would take the instructions to the ribosomes. The ribosomes would make the protein dictated by the gene. In other words, it might be possible to engineer bacteria to make human proteins.

After the Supreme Court ruled that man-made life forms could be patented, a young employee of Kleiner-Perkins, Robert Swanson, realized that armed with patents and gene splicing, you could create a new drug company by engineering bacteria to spit out crucial human proteins like insulin. So in 1976, Swanson talked Boyer into joining forces. At first, the company was called Herbob and consisted of Swanson and an empty office in downtown San Francisco. But when Swanson incorporated in April 1976, he needed a real name. Boyer suggested a derivation of "genetic engineering technology": Genentech.

Not only was Boyer criticized by his science peers for selling out, some biologists doubted whether the technology was anywhere near ready to make human proteins. Nobody had ever engineered a bacteria to accept human genes. Nobody was sure that, even if you *could* engineer such a bacteria, you'd be able to harvest the proteins that it made. In fact, Swanson had approached Cetus about using the technology and Cetus rebuffed him, arguing that it just would not work. But a few months after Genentech incorporated, Swanson announced that the process did work. Berg himself called the achievement "astonishing." The president of the National Academy of Sciences said it was "a scientific triumph of the first order."

Inco and Kleiner-Perkins both put money into Genentech. With that money Swanson built a small lab space in South San Francisco and began looking for new scientists. Dennis Kleid, Haseltine's former MIT collaborator, topped the list of prospects. Shortly after their collaboration, Kleid had left MIT and moved west to work at the Stanford Research Institute in Menlo Park. He and Haseltine kept in touch, and then, in February 1978, they met up at a scientific meeting in Colorado. Kleid told Haseltine that he and his postdoc, David Goedell, were being recruited. Haseltine "discouraged us from going to Genentech. He said we should stay with academia, that the company stuff was not a good idea," Kleid recalled.

"I told him, 'What the hell do you want to do that for, Dennis?'" Haseltine said.

"He thought we were nuts," said Kleid. Goedell and Kleid joined anyway, becoming Genentech employees numbers three and four.

On September 6, 1978, Genentech announced that Goedell, Kleid, and scientists from the City of Hope research center in Duarte, California, had harvested human insulin from *E. coli* bacteria.

Within months, everybody who was anybody was joining companies. Though it had money in Genentech, in 1978 Inco held a series of meetings in its Toronto offices to create a rival with Walter Gilbert as the headlining science star. It was named after the magic substance that gave life to the robots in the Capek play *R.U.R.*: Biogen. In 1980, another of Haseltine's mentors, Mark Ptashne, established yet another company.

"That made a huge splash on everybody's consciousness," Haseltine recalled. "First of all, here are my mentors, Wally Gilbert and Mark Ptashne. Second of all, they were fighting with each other, as usual . . . This caused an enormous amount of attention because here were the two stars of molecular biology at the time."

In 1979, the government had decreed that retirement funds were now allowed to invest in venture capital (VC) outfits. The VC business was suddenly flooded with hundreds of millions of dollars and the VCs, in turn, went looking for new places to put that money. All the elements were now in place for "biomania."

The dam burst on October 14, 1980, the day Genentech went public. People in biotechnology remember that day the way Americans can recall precisely how and where they first heard about the terrorist attacks on September 11, 2001, or that John F. Kennedy had been shot. When I asked, Haseltine related every detail of a trip he had taken to East Lansing, Michigan.

He had flown to the Michigan State University for two days of talks and meetings on his retrovirus work and on DNA damage and repair. At the time, it was standard procedure for guests like Haseltine to be paid a small honorarium of about $100 per day, air fare, and a motel room. The meeting schedules were often demanding: a long series of face-to-face encounters with colleagues, and group talks from morning to night. On his second and final day there, Haseltine did not have time for lunch. Then he had to rush to the airport to catch a plane to Detroit in order to make a Boston connection. He asked his driver to stop so he could grab something to eat. "We stopped at this grungy bar and I got something called a French dip." Haseltine was unfamiliar with French dips, roast beef sandwiches meant to be dunked into au jus. "I mean, you dip these things into grease! But I was starving so I dipped my roast beef sandwich into grease with the predictable result that about half an hour later I was really pretty sick. So here I was sitting at this airport, this tiny little airport, getting sicker and sicker. I picked up a newspaper and I saw that Genentech had just gone public."

Genentech was set to sell one million shares at $35 per share. The company had a terrific "story"—that genetic engineering would treat and maybe even cure any number of diseases—and the market ate it up. Before the day ended, Genentech hit $89. The power of that run shook Haseltine, turning him from fight-the-system corporate skeptic to enthusiastic, if slightly nauseous, capitalist. "I knew my friend Dennis had just made $2 million or something and I thought to myself, Ugh. $200 and a French dip. Maybe it's time to think about [starting a company]. If these guys can do it, I can too."

Swept up in the gene splicing frenzy, lots of people had the same thought. For example, Robert Foman, the president of E. F. Hutton, spent millions trying to get Hutton involved in biotech start-ups until, after seeing a UCSF scientist named John Baxter on the cover of the *New York Times* magazine, Foman decided to create a start-up around him. In December 1981, Foman called Al Scheid, a close friend and colleague who had gone into the wine grape business, and asked Scheid to set up the company and run it. Foman didn't care if the company had a product or not. "I said, 'Bob, they've got no drugs. They've got ideas. They've got research people. I do not want to do it,'" Scheid said. Instead he provided Foman a profile of the type of leader he thought the company would need: an entrepreneurial salesman. "The next day, a check arrived made out to me for $120,000 with a note saying, 'Start the fucking company.'" Scheid accepted the "offer" and slapped a generic name on the company— California Biotech—and within a year raised $32 million from individuals and institutions like Inco and Citibank.

Scheid admitted that in the post-Genentech environment he did not have to work too hard. "This industry was hot. The *Wall Street Journal* could not talk enough about it. We just told people the truth. This was a limited partnership for biotech research. This way people could have direct participation [in science]. I told them that I was only asking people who could afford to lose all their money and if they could not afford it, they should not invest. That's the art of sales, to make it exclusive, to make them want the product."

When the product is the future, everybody wants it. Investors were so fascinated by the potential of DNA that after a few cursory lessons on how the double helix worked, they were sold. "We made people insiders," Scheid said. Things got so wild during biomania that, Scheid recalled, "companies got started based on scientists walking out of labs with test tubes in their pockets."

At almost the same moment Haseltine was sitting in the Lansing airport trying to avoid becoming reacquainted with his French dip, a twenty-four-year-old aspiring songwriter in New York named David Blech watched the news of Genentech's IPO, turned to his stockbroker father, and said he wanted to get into the biotech business, too. Out of the blue, and with no science—and little business—background, Blech called a noted virologist, one of Haseltine's former collaborators, simply because he had read an article about him. Within days, Blech had signed up the scientist, named the company Genetic Systems because it sounded like Genentech, and began to pump for investors. Blech raised millions and suddenly had his own investment firm, D. Blech and Company.

Haseltine finally did start his own company, but he was slightly more cautious. A few months after the Genentech IPO, Haseltine flew west to give a talk at the Scripps Research Institute in San Diego and used the opportunity to travel up the coast to visit with Kleid, take a tour of the Genentech lab, research how companies were created. When it was time to catch his flight back east, he called a friend in New York, Ed Goodman, the grandson of a Bergdorf Goodman founder, who worked as an investment banker. "I said, 'Eddie, what kind of banker are you?' And he said, 'I'm an investment banker.' I said, 'Is that related to a venture capitalist?'"

Haseltine may have been naive about venture capital, but he had called the right man. Goodman worked at a venture firm until 1980, when he joined the Hambros Bank's venture group. So he was well connected in the world of investment. Goodman agreed to meet with Haseltine, mainly as a friend helping a friend, and Haseltine switched his ticket's destination from Boston to New York.

Goodman provided some basic tutoring, then handed Haseltine off to a colleague in the field. Haseltine, who never does anything halfway, became an avid student, attacking business the way he attacked every subject. He bought books, he read the financial pages, he asked questions. Within a year, he had an idea of how to make a company out of his knowledge of retroviruses. But what would Harvard say? The notion of professors starting their own companies was controversial. There had already been negative reaction on campus to commercializing biology. MIT and Harvard had been littered with "biobucks," sneering fake dollar bills accusing science of selling out, and Haseltine was just an assistant professor.

So Haseltine assembled a group of high-powered scientists to join him, including Max Essex, a tenured professor at the Harvard School of Public Health. If Harvard was going to object, it would have to fight an all-star cast.

Feline leukemia virus is a retrovirus that is one of the biggest killers of cats. Vets were desperate to find a way to prevent it. That seemed like a decent market if a company could come up with a product. As leader of the group, Haseltine took it upon himself to seek cash. He called Blech. Investing in what was now called Cambridge Bioscience was a no brainer for D. Blech and Company. It didn't matter what Haseltine and his colleagues were going to try to do, the science names were the big draw. So, with $2 million in initial funds, Cambridge Bioscience opened for business in Hopkinton, Massachusetts. It later moved to a new biotech park set up by developers in Worcester.

The company did eventually develop a vaccine for feline leukemia virus. But, according to Blech, the real goal was never animal vaccines, but human disease. "They thought there was a relationship between feline leukemia virus and regular human leukemia. They were not really interested in animal vaccines except as a vehicle for research into humans."

It turned out that Bob Gallo was right. Retroviruses did cause human disease. In 1979, a Gallo team isolated a virus dubbed HTLV 1 for human T-cell leukemia virus. In 1980, Japanese scientists discovered that HTLV 1 caused a human leukemia. Then Gallo attended a meeting at which a scientist from the Centers for Disease Control suggested a virus could be involved in a mysterious disease that was gutting the immune systems of gay men and, Gallo recalled in a *Science* memoir, "Max Essex reminded us that the feline leukemia retrovirus not only causes leukemia, but that its variants could also cause immune disorders." By the time the world was given the news that an HTLV-related virus, today called HIV, was the cause of AIDs, Cambridge Bioscience "had some ideas," Blech said.

Haseltine's business and his Dana-Farber lab plunged into the middle of the biggest politico-scientific storm of the twentieth century. The entire field of AIDs research was hounded by controversy from the start. A long saga of jealousies, accusations of deceit among rival scientists, street activism among homosexuals, and the devastating effects of AIDs on high-profile communities like show business and the arts created a toxic cloud. Within the government labs, the rivalries spilled out of control. Yet, for these very same reasons, AIDs research also became the single most glamorous field in biology. Jonas Salk was a legend, not to mention wealthy, for saving the world from polio. Scientists thought a new Salk would be created out of the AIDs pandemonium and, given the politics in virology, there was no shortage of aggressive candidates vying for the title.

A string of HIV-related announcements came from Haseltine, particularly on genes that controlled the virus. In April 1984, Cambridge Bioscience announced that it had filed patents on a test for HTLV III, then the U.S. name for HIV. Not only could it test for the virus, the company claimed, but the feline leukemia virus vaccine it had developed might wind up being useful against HTLV III. Cambridge Bioscience went public. Then, in 1988, it was granted the exclusive license to a protein found on the virus's coating that had been characterized by an Essex-led team. The protein, called gp120, proved critical to the company's most successful product, a quick test that used gene splicing technology.

Cambridge Bioscience made headlines, but repeatedly stumbled. Its quickie test, dubbed Recombigen, "the five-minute AIDs test," or, as Haseltine called it "the back-seat-of-the-car" test, was approved by the FDA on December 13, 1988, but the company never found its footing. Haseltine blamed the CEO. Blech blamed Haseltine and the other scientists. "None of these eminent scientists was willing at that point in their lives to devote their full time to this effort," he said. "I probed the advisers [Haseltine, Essex, and the others] several times to see if any would go full-time and I always got a negative response.

They were worried about their academic careers. They wanted to win Nobel Prizes, all of them."

Haseltine sold out in frustration in 1989. Subsequently, the company became embroiled in lawsuits, bought out another company, changed its name to Cambridge Biotech, and finally collapsed in 1994 amid a scandal over bogus sales and tricky accounting. The French company bioMerieux Vitek wound up with its assets.

After starting a score of biotechs, turning D. Blech and Company into a stock trading powerhouse, and amassing a fortune he estimates to have been in the neighborhood of $300 million, David Blech's empire came crashing down in 1994 amid a scandal over phoney stock trades. Panic roared through the NASDAQ, where most biotechs traded. After cooperating in a SEC investigation, Blech pled guilty to two counts of fraud and was sentenced to five years' probation. The first wave of biomania was over.

But AIDs made Haseltine famous outside the world of science cognoscenti. His name began showing up in newspapers around the world, his face on television screens. He became familiar with European airports and convention halls and began to circulate with a new, glittering crowd, high flyers in commerce, media, and show business who gave wonderful parties and raised money for AIDs charities. He began earning a reputation around Harvard as a sport. For years he had driven an old, beat-up convertible with a malfunctioning top, but colleagues now saw him driving up to Dana-Farber in a new Mercedes ragtop, a jaunty fedora perched on his head.

WALLACE STEINBERG *believed.* He believed in himself, in science, in the power of money. Most of all, he believed in immortality. If he had not believed, he could never have sold so many other people, including Haseltine, on all these. But he did believe and he managed to convince himself that if he could put all the ingredients together just so, he stood a very good chance of living forever. The science of the previous fifty years had set the directional arrows. In the wake of biomania Wally Steinberg reckoned he had found the way. Venture capital and high-profile scientists were the key. Stir them into a pot, give them the freedom to simmer the nutty, outlandish ideas that the government grantors would not fund, keep the alchemy going long enough, and then let the magic pour. Steinberg knew that the real lesson from biomania was not whether you could actually do what you said you wanted to do, but whether you could convince others, preferably those with a lot of money, to buy into the dream.

Genentech was setting its sights too low. Steinberg did not just want to start companies that might make products to treat disease, he wanted to cure disease, permanently. "I have this theory that death is a genetic disease," he told *New York Times* reporter Gina Kolata in 1992. "There is no religious, preordained reason to die."

"After a meeting, Wally took me aside," recalled biotech entrepreneur and scientist Michael West, of an early conference he had with Steinberg. "He took me into his office and closed the door. He said he had been the vice-president of planning at Johnson and Johnson and was asked to do a strategic plan for the future of health care at J and J. He did it, and presented it to the board of directors. He said he walked into the board meeting, and his opening line was 'Gentlemen, Johnson and Johnson's mission will be to conquer death!' I said, 'Well, what did they say?' He said, 'Well, after a short period of discussion, they fired me.'"

Steinberg's tale to West was almost certainly apocryphal in its details, but Steinberg's interest in life extension was not. Steinberg would start conversations with scientists by telling them he wanted to cure death, and he did not mean the statement to be hyperbolic. "He would tell any scientist he could that he was going to find the secret to let him live a couple hundred years," former Nixon (and later Gerald Ford) aide James Cavanaugh, who became Steinberg's business partner, said.

That would sound ridiculous to most people, but not to Steinberg and not to his friend and frequent tennis opponent, investment banker Billy Walters. Steinberg's certainty was unshakable. "He totally believed in himself," Walters said. When the two played on the weekends at Steinberg's mansion in New Jersey, Steinberg always thought he could win. He rarely did, losing cash pots "into the five figures. He thought he was a better tennis player than me and every weekend I'd go out to his house and take money from him, but it never deterred him from playing." Walters appreciated Steinberg's determination because Walters and Steinberg also shared an interest in life extension. Walters was an enthusiast of offshore longevity clinics and a chronic user of HGH, and Steinberg believed that growth hormone just might boost him over the hump into immortality.

Steinberg was a mass of a man, tall, physically imposing, with a protruding gut and an equally outsized manner who had grown up in Brooklyn's poor Jewish neighborhoods. After he graduated from Rutgers University with degrees in pharmacology and pharmaceutical chemistry, he went to work for a lesser known drug company, then moved on to Johnson and Johnson. He worked his way up through the ranks, becoming famous within the company, and in the home care products industry, by directing the creation of the Reach toothbrush.

The Reach was more concept than new product. It was bent, something like a dental tool, so that, Steinberg argued, it would remove plaque like a dental tool. The shape did not really impart any great advantage over other toothbrushes, but Steinberg knew that tangible benefits were not really the point as long as people believed and were willing to pay a premium for a bent toothbrush. "He got the company to spend on TV ads the way it never had before and he got people to spend money on a toothbrush like they never had before," Cavanaugh marveled. "Wally was a super salesman."

When the company decided to form Johnson and Johnson Development Corporation, an internal venture capital operation to fund initial-stage development of new products, Steinberg and fellow employee Hal Werner were tapped to lead it. The position enabled Steinberg to witness the birth of the new biotechnologies and he came to realize that the first to move into them might be able to set an agenda for an entire industry.

Steinberg left Johnson and Johnson and briefly joined an old-line, white shoe venture firm in New York. It wasn't a good cultural fit. Steinberg was not a white shoe type. He was brash, aggressive, rough around the edges. "Everything revolved around him," said Billy Walters. "He was the kind of guy who could not believe that penicillin was discovered without him." So Steinberg left and eventually formed Healthcare Ventures, a venture capital outfit, with Hal Werner, his former Johnson and Johnson colleague, and James Cavanaugh.

With his pharmacy degrees, Steinberg imagined himself something of a scientist. He was convinced that an undercurrent within biology, if nurtured with the right combination of promotion, salesmanship, and research, could usher in a new age of human transformation that would dwarf the first wave of biomania. By the mid-1980s, the first wave was simmering down as some companies, floated on the hot air of the moment, began to fall back to earth. Others were absorbed into bigger pharmaceutical firms or morphed into other entities; a few, like Genentech and Biogen, survived. In the hyperspeed way technology becomes old news, the idea that you could make human proteins by genetically engineering bacteria or some other cell was no longer revolutionary. And only a few of those proteins, like insulin, HGH, and interferon, were actually being made. They were no panaceas either. After all, you could give a diabetic all the recombinant human insulin you wanted, but the patient was still a diabetic.

Steinberg reckoned there had to be a next step. Whatever happened to the exciting work of the 1960s and 1970s? Out of all those *E. coli* and viruses and cells that had been manipulated and those genes that had been discovered, a new human biology had to be possible. Somewhere in the academic labs and in the NIH, there had to be scientists willing to put that theory to the test, willing to be as daring as he was, by applying the earliest, most basic science and turning it into medicine.

Steinberg was a man of great ambition. He wanted to make money, lots and lots of money, which of course would be a necessity if he was to become immortal, but he also craved wealth as a way to show up the Manhattan swells he used to envy growing up across the East River. Venture capital had been dominated by the family fortunes like that of the Whitneys and the Rockefellers and Steinberg liked the idea of seizing the opportunity to get in on a new industry they had not yet discovered, elbowing his way to the cash trough, and securing his place with the old money. When Healthcare Ventures would hold meetings in Manhattan's Harvard Club, where Werner, a Harvard MBA, held a membership, Steinberg would perform a schtick, looking around at the leather chairs and paneled walls and remarking to the company's guests how incredible it was that a poor Jewish kid from Brooklyn would wind up lunching at the bastion of elite Waspiness.

The way to make that money, Steinberg thought, was to find the ideas before the other guy. He used to say that he wanted to "own American science," at least the biomedical sciences, and the only way he could do that was to lock up new technologies. If it was new, so new that nobody else believed it was possible, that was the idea Steinberg wanted. "He could be dark and cruel," Nancy Howar, a former business associate of Steinberg's, said, "but he was visionary." Everybody thought so, even the people—and there were many—who did not like him. "There were no limits to what we could dream," recalled Mona Geller, one of Steinberg's early associates. "We were not told to think of limits. Our mission was to think without limits."

Scientists who later met Steinberg weren't quite sure what to make of his thirst for early-stage science. They describe him as the kind of larger-than-life character you might imagine standing with his thumbs hooked around his suspenders explaining why you'd be crazy, absolutely insane, not to buy this here swamp. Why, you do a little backfill, plant a few trees, and you've got Palm Beach.

"Wally was a first-class crook," Florence Haseltine said, alluding not to criminal activity but to accusations from some scientists that Steinberg was willing to fudge scientific data and finesse the details of agreements to get deals done. "But," she hastened to add, "he was also one of the most human people I've ever known."

"I do not know about the crook part, but I would say he was definitely a competitor," Jay Short, CEO of Diversa, a company founded by Healthcare Ventures, said. "Wally was willing to sell the dream and the details were not the issue."

But the long tradition of scientific skepticism, of not exaggerating one's research results, and the possibilities of scientific breakthroughs make scientists a skittish bunch when it comes to business and claims made for their work. A big, blustering salesman like Steinberg, one who tended to open conversations by talking about life extension, had very little access to the science tribe. He knew the drug business. He knew how to make an investment and how to trade stocks, but he had a short track record in venture capital and almost no track record when it came to biotechnology. When he set out on his own from Johnson and Johnson, he hoped to follow the path paved by Alafi, Inco, and Kleiner-Perkins, but he did not know how to attract a Walter Gilbert, or a Bill Haseltine.

Deeda Blair did. She would prove to be the vital link between Steinberg, Haseltine, and the popularization of the idea that human beings really could direct their own evolution.

SMOKE FROM A THIN CIGARETTE curls up from Deeda Blair's fingertips, slowly spreads like a veil in front of her green eyes, then wafts past the trademark shock of white hair within a sea of black that swoops up off her forehead. She's wearing a skirt and a wool knit blouse over a frame that has not gained a pound since the 1950s when she was a society ingénue who walked into the atelier of an up-and-coming Paris designer named Hubert Givenchy, tried on clothes made for his models, and found they fit. She is seventy-one years old now, and hates the fact I know it. But she is still beautiful. She has the polish of a diamond. Her voice is tinged with a slight rasp from years of smoking and she speaks so softly the effect is seductive. I cannot imagine a man refusing any request she might make.

We are sitting in her living room inside the large federalist-style home she shares with her husband, William McCormick Blair Jr., in the Wesley Heights section of Washington, D.C. Photographs of the Blairs with John F. Kennedy, Adlai Stevenson, and social friends sit on tables and bookcases. Big coffeetable books, compilations of *Vogue* photography and art books, are laying on a table and the bookcases are filled with histories and politics and more art, all a reflection of the life she has led.

Deeda Blair was born and raised in a prominent Chicago-area family, the daughter of an attorney, and educated by the Sisters of the Sacred Heart in a convent school. About the only formal science education she ever had was a little physics, taught by nuns who were part of a Church that had been suspicious of science since before Galileo. She lived in Paris, for a time, a refuge after her brief and unhappy first marriage into another prominent Chicago family. There, she became a devotee of haute couture. She made the perfect client. She was rail thin, elegant, stunning, a less famous version of Audrey Hepburn (who was discovering Givenchy at about the same time Deeda Blair was). She also purchased Chanel and remains a fan of the Chanel suit, and of couture in general, to this day. Several weeks before I met her, I saw her on TV attending the Paris debut of American designer Ralph Rucci. She is sometimes pictured in *Vogue* or *W,* sharing pages with the vaguely titled like Marie Chantal of Greece, certain Middle Eastern women, and bright young things like the Hilton sisters—all the beautiful, rich people young and old who seem to live in a parallel universe of charity balls and European night clubs.

Deeda was a perfect match for William McCormick Blair Jr. Blair, a descendant of the inventor of the McCormick reaper, a son of William Blair, the founder of what became one of biggest investment companies in the Midwest, was the aide-de-camp to Adlai Stevenson, both during Stevenson's governorship of Illinois and throughout two campaigns for president of the United States. After he and Deeda married in 1961, he was made an ambassador, doing one stint in Denmark and one in the Philippines. The Blairs became an international couple, examples of American style, wit, and charm. As part of Washington's power elite, it was virtually ordained that Deeda Blair eventually fall in with Mary Lasker, Florence Mahoney, and the world of biomedical boosterism.

In 1942, Mary Lasker and her husband Albert, an advertising mogul who was once the majority owner of the Chicago Cubs, created the Albert and Mary Lasker Foundation for medical research. In addition to becoming a powerful advocate of the NIH, the foundation began awarding an annual prize for innovative research. Since its inception in 1944, the Lasker Award has become the single most prestigious biomedical prize in the world aside from the Nobel. Indeed, it is often referred to as "the American Nobel," and it is a frequent bellwether to future Nobel winners. Greased by old Chicago ties, Deeda Blair and Mary Lasker became close, and, in 1965, Deeda was named a director and vice-president of the Lasker Foundation.

Through Mary Lasker, Deeda Blair also became friendly with Florence Mahoney, a Washington, D.C. socialite, health lobbyist, and former patient at the antiaging Clinique La Prairie. Florence Mahoney and Mary Lasker were a formidable Washington duo who tirelessly used their social and political connections to advance the cause of medical research. Mahoney's dedication to aging research, combined with her powerful influence, is often credited with the creation of the National Institute on Aging within the NIH. She hoped the agency would seek answers to the questions at the heart of aging itself, the "why" of the aging process, and ways to slow it down.

Through her association with Mary Lasker, Deeda Blair's influence became enormous, too. She was appointed to boards and committees affiliated with organizations from Duke University's Comprehensive Cancer Center to the Harvard School of Public Health, to

Rockefeller University to the American Cancer Society. At one time or another she has sat on some of the most prestigious boards and committees in the world of biological research. Yet she is not a scientist, or a politician, or a businesswoman. Deeda Blair is a catalyst, the hostess of an extended salon who is expert at making the kinds of introductions that keep the party going, a talent that makes others want to know her.

"We used to do large dinners downstairs," she says, indicating the lower floor of her home, "with a mix of scientists and congressmen and press people. A friend of mine once walked down those stairs and said, 'My gosh, you've got the whole appropriations committee here!'"

She places her elbows on her knees, cocks her wrist so the cigarette between her fingers is aimed out a window, and says, "I can pick up the phone and say this is Deeda Blair, I am vice-president of the Alfred and Mary Lasker Foundation and I am a member of the research committee of the American Cancer Society and usually people will see me." We look at each other a moment and then laugh at the understatement.

When Mary Lasker wanted to mount a nationwide blood pressure awareness campaign, she called Deeda "and said do you know anybody who knows Elliot Richardson?" then secretary of health, education, and welfare under Richard Nixon. "I said, 'Well, two nights ago there was a party at the British Embassy and I spent a large part of the evening waltzing between the columns with him.' She said, 'Do you think you can get an appointment?'" She smiles at me like a sly sorority coquette. "I said, 'Yeah, I think I can.'"

Within weeks, Mary Lasker had the money she needed to start the campaign and Emerson Foote, of the giant ad agency Foote, Cone and Belding, the agency started out of the remnants of Albert Lasker's old firm, arranged free nationwide ads.

Politicians and businessmen are not the only ones Deeda Blair can bend. Scientists, in their platonic way, have fallen for her, too. Deeda and Mary Lasker spent years touring the hottest sites in the biology landscape in both the United States and Europe, and, among others, Deeda became good friends with Bob Gallo and Max Essex. Once biotech became a force, she and Mary Lasker learned as much as they could about that, too.

They toured Genentech when it had just a handful of employees. They went to Sweden's Karolinksa Institute to see the earliest investigators of a new anticancer drug, the human protein interferon, and then, in Cambridge, Massachusetts, they stopped by to see Wally Gilbert. He was just starting to create Biogen with Inco and they hoped Biogen's first drug would be interferon. Deeda requested a private tutorial and Gilbert gave it to her. When he told her he was planning to work on interferon for Biogen, she explained her already deep knowledge of the subject. "He looked at me and said, 'Does it work?'"

Because she could talk science with the best and hold her own, because of her position with the Lasker Foundation, and because she has always believed in the power of science, scientists trusted her. She repaid that trust many times over. She is still the one person in the world who can sit next to the catwalks of Paris during the spring shows, put the arm on a few of the 500 or so women who regularly support couture by purchasing dresses priced about the same as a BMW, and come back with quarter of a million dollars for a new DNA sequencing machine for somebody's lab. "Deeda could walk through the NIH and everybody would genuflect," an acquaintance told me.

Deeda Blair and Wallace Steinberg first met when he was at Johnson and Johnson. She knew many pharmaceutical company executives and back then Steinberg was just another one. The thought of becoming involved in the business end of biology, of actually helping to finance companies, had never occurred to her.

But then a friend, Nancy Howar, suggested that they form a two-person art consulting business. Howar knew important interior designers; Deeda Blair had always been an international tastemaker. Combined they might be able to make a nice bit of money helping designers choose art and antiques. They formed Wesley Associates, a discreet name derived from their Washington neighborhood, and soon they were selecting the valuables to adorn the newly decorated offices of elite corporate clients like American Express and R.J.R. Nabisco. But Deeda never neglected her beloved scientists.

"I would hear Deeda complain that grants were drying up and these world-class scientists could not move their work forward," Howar recalled. So Howar suggested that Wesley Associates move into biotech the same way it had moved into art consulting. Maybe they could help link money to the biologists who needed it. With help from her husband's law partner, Howar and Deeda Blair met with a couple of up-and-coming venture capitalists. Wally Steinberg was one of the first.

Over lunch in New York in late 1985, Steinberg told Howar and Blair that he wanted Wesley Associates to connect him to the best scientists in the world. From Deeda Blair's contacts, he hoped to create the greatest scientific advisory board venture capital had ever seen, a board that would give him and his company instant credibility.

Deeda was skeptical at first, of both Steinberg and the concept of playing matchmaker. But she agreed to think it over during the Blairs' annual vacation in Venice. Then, the day after she returned to Washington, she attended a meeting at the National Cancer Institute where it was announced that, for the first time, NIH scientists would be permitted to consult for industry. Suddenly the field of candidates was broadened. Feeling blessed by the government itself, she decided to go ahead and, just three weeks later, Wally Steinberg had his board.

She tried to recruit Haseltine. He and Deeda Blair had first met the year before at a small but prestigious AIDs meeting in Paris and had struck up a fast friendship. Haseltine knew a good connection when he saw it, and he and Blair began a mating dance. "He was very interesting. We talked science, but he also wanted to go to a tea shop and get fantastic pastry, or we'd stop at a florist and he would buy me one white lilac. And later he'd send me a Henry James book."

She especially admired the fact that, like her, he was in a hurry to translate cutting-edge science into clinical practice. She had another interest in him, too. She wanted to place Gallo and Essex in line to win a Lasker in recognition for their work in AIDs, and she wanted Haseltine's help. He was eager to contribute, even if he was slightly peeved. "In retrospect, I had as much a role as they had, so I should have been on the list of recipients rather than the list of people helping behind the scenes to get them the prize, but I sort of got snookered on that one. But they did get the prize and I worked mightily on their behalf."

She introduced Haseltine to Steinberg in an effort to put him on the scientific advisory board, but despite her best efforts, he refused to join. Steinberg was not his type of character. "I found him obnoxious, self-important, crude," Haseltine recalled. "I told Deeda 'You shouldn't work for him and I shouldn't work for him. He is an unpleasant person.' That was a mistake and it took me a year and a half to remedy that mistake." Finally, "Deeda said, 'You ought to meet with him again.' I found him to be more reasonable," even if "he had not changed any of his other characteristics." Haseltine signed on.

Even with the gold-plated advisory board, Deeda Blair and Wesley Associates proved invaluable for his success. Fourteen years after Theodore Friedmann and Richard Roblin proposed gene therapy, Steinberg wanted to make a company out of it, despite the fact that fixing malcontent genes, or putting new genes into a human being to make up for some deficit, was still just a theory. So Deeda Blair swung into action, making up a list of all the top people, including Inder Verma and NIH scientist W. French Anderson. Then she played the trustworthy liaison between Steinberg's company and scientists.

"Hal [Werner] and I went to see them," she recalled. "French was very leery of commercialization and Healthcare could not get an appointment. But I could call up and say, 'Look, you have nothing to lose by just meeting with them. I see this as a way of moving and expanding the science in your lab. Let's just have a meeting.'" They did and the world's first gene therapy company was created out of Anderson's NIH experiments.

Deeda became the face of Healthcare Ventures in the world of biology. She was the guarantor of Steinberg's bone fides, wooing scientists into meetings, holding their hands when they expressed reservations about Steinberg or the deals they were making. She was part of the Lasker, after all, and all those other boards and committees, perhaps the most cross-connected person in the world of biomedical research, and she was speaking up for Wally Steinberg. Once they met with Steinberg, scientists were lured by two other powerful attractions: money and freedom. Steinberg gave away cars, generous stock deals, and promised the scientists, most of whom came from the NIH at first, they would get rich. It was all a far cry from their government salaries. Best of all, he promised their ideas could come true. No more bureaucracy, no more grant applications, no more justification to review committees.

Every time another company was funded, Wesley Associates got a piece of the action in the form of founders stock. But eventually, Steinberg began to wonder why he needed Howar. In November

1989, Steinberg took Howar and Blair to lunch at New York's fashionable Lotos Club and Howar was told that her services would no longer be required. The Wesley partnership—and her friendship with Deeda Blair—was disbanded.

With the mining of NIH science well along, Healthcare Ventures turned to Harvard. Once again, in return for a cut, Deeda ran interference, assuring skeptical administrators and scientists that Steinberg could, and would, pull off commercialization of the biology going on in the labs. He wanted to start a dozen or so companies, he said, deals that could plow perhaps $100 million into Harvard science in return for intellectual property.

Sure, the science was delivered from the womb of the NIH and the universities prematurely. But Steinberg thrived on the sizzle and the big story. "Nobody else would have funded some of these companies," a former Healthcare Ventures employee told me. "They were early and pioneering in terms of their impact and we dreamed them up." Gene therapy, for example, was hardly ready to commercialize. Almost twenty years after the formation of Genetic Therapy from French Anderson's NIH experiments, the field is still considered clinical research. But how exciting it was to think of tinkering with people's genes to fix their ills! Or creating a company to engineer animal organs so they could be transplanted into people, or engineering human cells!

According to his own definition of success—$1 billion in annual sales—not one of Steinberg's companies has been successful. Some have been purchased by bigger companies. Genetic Therapy, for example, was bought by Novartis. Yet by another measure, Steinberg was wildly successful. He sold hope and optimism and people bought it. Steinberg's dream, and Deeda Blair's, too, as well as the dreams of the scientists they recruited, helped set the agenda for much of biology and biotechnology. "Technologies would have gone by the wayside," the former Steinberg employee said. "Nobody would have had the sheer acumen to develop companies that had these impacts on society." The impacts were not always practical. They just had to fire imaginations with the idea that biotechnology could, and should, aim high, that the ultimate solutions to disease, death, and human improvement were out there, somewhere, if only somebody had the guts to breathe life into them.

In no instance was this truer than in the formation of Haseltine's own company, Human Genome Sciences. When the international Human Genome Project kicked off in October 1990 with James Watson at the helm, it was compared to the space program. Causes for all sorts of diseases, not just cancer, would be found. Then those diseases could be ripped out by their genetic roots, not just treated with drugs. It was the triumph of the Schrodinger-inspired reductionist biology. But the project rapidly hit quicksand. Not only was the technology slow, but sequencing the genome was not really a good way to find genes, which make up only about 3 percent of the entire genome.

Some in science, Haseltine and Venter among them, dismissed mass genome sequencing as a wasted effort. "At that time, Craig used to say, 'You'd have to be a complete moron to want to sequence the human genome. Everybody knows it's about the genes,'" Haseltine recalled. But James Watson was running the genome project and Watson did not want to use Venter's technology because cherry picking genes, as opposed to brute force sequencing of the whole genome, would lead to patents on medically useful gene sequences. Watson opposed the idea of the government taking out patents resulting from the project, a stance Haseltine called "off the wall nutty." Watson's stubbornness on the issue left the door open for Steinberg and those who followed him. "If they had [used Venter's system] as part of the Human Genome Project, then Human Genome Sciences or this whole genomics industry would not really be here," Haseltine said. "It just would not have started."

But the project did not accept Venter's methods and he felt jilted. Once again, Deeda Blair played an important role when she and Hal Werner visited with him after another venture capitalist told Cavanaugh that Venter was ripe for picking. "I was familiar with so much of the instrumentation he was using from all these biotech trips," she said, "and I watched, fascinated with how much of it had been roboticized. Then we went out for lunch and Hal had to catch a plane. I stayed, chatting with Craig, and found myself in kind of a slightly superficial conversation and I said, 'What are your hopes and dreams?' I don't say that sort of thing to people, I'm not that revealing, but I said it, and he said, 'Well, you can answer that. Will I ever get an institute at the NIH?' I said, 'No, I do not think you will.

There are already too many institutes.' He said, 'Well, I have an alternative with Amgen for $80 million.'" Venter told Blair he had three weeks to make up his mind about whether to hitch his star to the biotechnology company. Then Blair "went home and called Wally."

In the years since Haseltine joined Steinberg's advisory panel, Steinberg had taken him under his wing. The two men became close. Haseltine had taken part in the formation of a number of Healthcare Venture companies as an adviser and consultant and his income from Harvard, his consulting fees, the stock he was being granted, was making him far wealthier than the average professor.

But Haseltine's marriage to MIT professor Patricia Gersick was breaking up. In 1989, during an AIDs benefit in the Hollywood home of a flamboyant art collector, Haseltine met Gale Hayman, the ex-wife of Giorgio's of Beverly Hills founder Fred Hayman. In a variation of the old line about etchings, Haseltine offered to explain the art in the home. Soon, the two were an item. On weekends, Haseltine would commute from Boston to Hayman's New York pied-à-terre and, while in the city, he and Hayman would have lunch with Steinberg and Steinberg's wife.

One Saturday, as the couples were strolling down Park Avenue to a little French restaurant, Steinberg told Haseltine to park the women at another table. "He said, 'This is serious today, Bill. I've got an idea that will get you out of Harvard,'" Haseltine recalled. Haseltine did not necessarily want out of Harvard. He was making a lot of money. After a long battle, he had finally won tenure. He was helping Healthcare Ventures start companies and he got 5 or 10 percent of each one. But as the two men sat in the restaurant, Steinberg insisted that Haseltine go see Venter and consider running Human Genome Sciences.

Haseltine was not Steinberg's only option. Steinberg interviewed at least one other candidate and Venter was known to prefer an executive named George Poste, the research director with the drug giant Smith-Kline Beecham, one of James Cavanaugh's alma maters after his stint working at the White House. But Steinberg wanted Haseltine.

Deeda Blair was not so sure. When Steinberg consulted her about choosing Haseltine, she took the weekend to think it over. After all, it was going to be a complicated deal. Because Venter was

leery of corporate science, something he would get over in a few
years, Steinberg had agreed to set up Venter in his very own
autonomous institute—The Institute for Genomics Research,
TIGR—for about the same amount of money Amgen had been talking about. Human Genome Sciences would enter into a ten-year
contract with TIGR to hunt genes. Then Human Genome Sciences
would use the sequences to make products or sell the information to
other companies. Steinberg and Healthcare Ventures never did want
to start a nonprofit institute like TIGR, "but," Diversa CEO Jay
Short said, "if that's what it took to figure out how to take genomics
to the next level, that is what they were going to do" despite the fact
that Haseltine and Venter "could not fit in the same room." Blair
worried that an experienced CEO, an MBA at least, might be better
suited to the task. But "Wally always understood people's capacity
and Bill's capacity was pretty vast."

Haseltine went down to Bethesda to meet with Venter. When he
was through, the first person he called after reporting in to Steinberg
was Craig Rosen, his old postdoc from Dana-Farber. "I am in D.C. at
the airport on my way back home," he told Rosen. "Tomorrow I want
you to come back and go see Craig Venter and set up a company."
Then he called Rosen's wife, who was just settling into a new home
the couple had purchased in New Jersey, and said, "Cindy, did you
know you are moving to Washington?"

The rest of the world did not make distinctions between genes
and genomes and sequencing and messenger RNA and DNA. Most
people outside science just knew that something big was happening
and they knew it because scientists, the government, and people like
Wally Steinberg kept telling them so. When the genome project was
considered, along with the enormous run-up in the budgets of the
NIH, the spread of biological research all over the world, the hopes
for gene therapy, and the increasing power of computers that could
aid scientific research of all kinds, it seemed obvious that the New
Age the transhumanists and the life extensionists had hoped for was
dawning, bringing longer lives, better health, miracle cures. "Enormous power has been inherited by humankind in this new biology,"
former NIH director Donald Fredrickson wrote of the new biotech
era, "offering myriad possibilities for improved health practices in
the future." Steinberg thought genomics was so important that he

considered the founding of Human Genome Sciences and TIGR to be in the national interest. The answers to many questions were in our genes, he argued, and the United States had to find them first.

Genomics became the big biotech buzz and Steinberg saw himself becoming the gene Rockefeller. He had a grand plan. There were to be three sister companies feeding off TIGR: Human Genome Sciences, Industrial Genome Sciences, and Plant Genome Sciences, a Standard Oil of genomes. If anybody wanted to use gene-based technologies in agriculture, industry, or human applications, Steinberg wanted them to have to come to a company Healthcare Ventures had founded. It was all part of his ambition to "own American science." Plant Genome Sciences never got beyond the sketch-on-a-napkin stage, but Industrial Genome Sciences did, and was eventually renamed Diversa.

HGS, though, was the star. Before the ink was dry on the incorporation papers, it became clear how eager big pharmaceutical companies were to embrace the future. The deal between HGS and TIGR was fragile from the start, not least because of what many insiders viewed as an inevitable clash between two ambitious, driven men like Venter and Haseltine. But as fragile as it was, the deal between TIGR and HGS was good enough to entice Smith-Kline, James Cavanaugh's old company, into paying for the richest collaborative agreement in drug company history, a $125 million deal. Smith-Kline was so anxious to seal the bargain that it sent George Poste to Florida to meet with Cavanaugh, who was attending a meeting there. Poste carried a $1 million check as earnest money.

Armed with the huge endorsement of Smith-Kline, Human Genome Sciences went public in December 1993. The stock market, venture capital, science, and the wide public all bought into the HGS story that it would be first to find and exploit many if not most of the protein signals that cells send to each other to carry out the functions of life. With those signals in its quiver, HGS would then be in a position to build the bridge Watson's Harvard biology program had wanted to construct between genes, proteins, cells, and tissues. The company founded by Wally Steinberg and Healthcare Ventures would have much of the the instruction manual for human beings.

Steinberg was the first to succeed in bridging the gap between the fantasies of H. G. Wells, and the early biology visionaries like

Haldane, and the new science mainstream. Schrodinger's imaginary clockwork was now, thought many, on the verge of being fully elucidated. A new industry, biotechnology, had been created on the backs of the molecular biologists Schrodinger had inspired and the money of the venture capitalists. And Wallace Steinberg proved that if you were willing to take big risks, you could make people believe in big rewards, the ones so many had long been waiting for. Few besides Steinberg and transhumanists and life extensionists talked about immortality or human enhancement out loud, but it was already the subtext of science and biotechnology. Steinberg wanted to make this effort more explicit. He was considering starting an antiaging company as early as 1992, telling the *New York Times* that he was looking for scientists to start one. But others would have to go the last mile. The larger-than-life Steinberg did not make it into the ranks of the immortal after all. He died at the tender age of sixty-one in July 1995. The official obituary said he suffered a heart attack in his sleep. According to an unofficial version, though, Wally Steinberg dozed off and choked to death on his false teeth.

6   Way Out West

Twice as many hopefuls showed up at the A4M conference in Las Vegas in 1995 as had the previous year, and, at first, they were treated to much the same message. There was talk of "phyto-chemicals," a laundry list of vitamins, a presentation on a hormone called DHEA, more discussion of HGH, shrill denunciations of the FDA, even a visit from the doctor who was then running Clinique La Prairie. He told the eager audience about a new and improved treatment called CLP, a French acronym for liver extracts of fetal sheep. But midway through the meeting, two new faces appeared, not apparitions of glowing white angels, exactly, more like a couple of middle-aged white guys standing on a stage. But the effect was the same as if Gabriel and a pal had swooped down from heaven because they represented something powerful and unprecedented in the world of antiaging and life extension. They represented science. Michael West and Calvin Harley identified themselves as leaders—Harley as the chief scientific officer and West as the founder—of a new California biotech company called Geron.

Harley, a somewhat dour, thin redhead, a former professor of biochemistry at Canada's McMaster University, a one-time Cold Spring Harbor researcher, author of a long string of scientific papers in prestigious journals, had come to introduce the life extensionists to the world of molecular biology. But he wasn't happy about it. This was all West's idea. His boss had to twist Harley's arm to speak at A4M, and, from my seat in the audience, Harley seemed very sorry he had come, looking, as he did, like a respectable accountant walking into Fat Lou's Midnight Pleasures bookstore in a town where everybody knows everybody else.

Harley's assignment from West was to deliver the good news that Geron, in collaboration with others, had isolated the messenger RNA for an enzyme called telomerase. With the RNA in hand, the company could then use the techniques developed twenty years before by wizards like Inder Verma to turn the RNA into DNA and then make copies of the telomerase gene itself. This was good news because, as the more devoted life extensionists already knew, scientists had a theory that telomerase could make cells immortal. It could keep them healthy rather than letting them sputter out or "senesce" like other cells do. If that were true, as it later turned out to be, well, then, you could make immortal tissues out of those cells, and from immortal tissues, it was just a hop, skip, and a jump to immortal people.

Clearly this was something different for the A4M crowd. Before Geron's appearance, life extensionists had only brushed up against molecular biology in the science fiction stories many of them read with devotion, in the HGH that was a product of the Cohen–Boyer experiments, the Genentech work, and the first wave of biomania. And for several years, there had been vague talk about "aging" genes in certain laboratory animals like fruit flies. But now Harley was telling them that mainstream science was being brought to bear on a problem that had always had a greasy snake oil film about it: life extension.

West shared none of Harley's discomfort. He was home and these were his people. He stood proudly on the stage, a tall, handsome man with a receding hairline, his smile betraying the pleasure he felt at delivering the prophecy that someday soon everybody could throw away the herbs and the potions and start eating like normal people again because real, honest-to-God science was on the case.

West had been on the case for fifteen years or so by the time he spoke in Las Vegas and his appearance was an improbable victory lap, the partial fulfillment of a vision he had while eating a hamburger in Niles, Michigan. It was around the time of his father's death, about 1980, when West took a lunch break from his job at West Motors, his father's small International Harvester dealership in Niles. As he sat eating his burger, he looked out the window and saw the town cemetery across the street. Like Paul on the road to Damascus, he was whacked by a revelation: everybody he knew, everybody he loved, would wind up in that ground someday. Then, boom! just like that, defiance welled up inside him, a determination that such a future could not be allowed to happen. Death had to be stopped. And if not by him, then who?

After all, science was his life. As a little kid, he set up a mini lab in his garage. Didn't matter what the particular subject was—biology, chemistry, electronics, physics—he liked them all because he had questions about how things, how everything, worked and he thought science could go a long way toward answering those questions. By the seventh grade he was waking up on school day mornings and peeling electrodes off his head. He had set up an oscilloscope in the attic and run wires down through the ceiling and attached them to leads on his skull so he could record his brainwaves overnight, roll the printed tape up in the morning, and study the electric output of his brain during school.

But science did not provide all the answers. What about the immortality of the soul? Science had precious little to say about that. Religion, on the other hand, claimed to know all about it. So West began an intense study of religion, eventually becoming a fundamentalist Christian. He learned Hebrew and Greek so he could read the Bible in languages closer to Jesus', proselytized to friends and neighbors. When he got a little older, he hung around bars handing out "Chick Tracts," miniature graphic comic books in which roughneck men and good-time girls dismiss Jesus, only to find themselves impaled by a steel beam, or taking a bullet to the head in a mugging, and staying conscious just long enough to see the grim, gnarled hand of death reaching out to escort them to hell.

West attended Rensselaer Polytechnic, a highly regarded science and engineering school in New York State, then enrolled at Andrews University in Battle Creek, Michigan, to study for a master's degree in biology. Andrews, a Seventh-Day Adventist institution, staunchly defends creationism, arguing that evolution is a lie and that the Christian god is responsible for making the universe. Soon after arriving at Andrews, West discovered that his fundamentalism clashed with his head for science. There was, after all, data—and West believed in data—that seemed to show that the earth was far older than the few thousand years the creationists had pegged from their reading of Genesis.

Gradually, West's faith disintegrated. He kept looking for some sign, for some tiny bit of evidence that a creator had left a fingerprint, someplace, any place, that would reveal the act of creation the way hammer strokes leave a dent in the wood next to a nail in a house, proof that somebody had built the thing. He came up dry. He looked into Buddhism, learned about the mystery religions of ancient Egypt, studied Greek mythology. "I have searched every corner of the edifice of religion and I think I know better than many people any basis for hope that death is not the end of existence, and I came up bankrupt."

Rather than believing in the literalness of religion, he now began to think that the rapture foretold by many religions, especially Christianity, was a metaphor. "If you go back and look at this, the apocalyptic hope, why would people build up belief systems that the world would come to this conflagration? In my opinion these millennial beliefs have their root in the recognition that the existential thing is correct: everybody we know will die, and all meaning becomes dust. That's no fun. So [believers] think this world and the evils we live with will undergo a climax and will change. We'll all be made immortal. The origins of apocalyptic belief are a hope for intervention, a rescue."

West, who is now fifty-one years old, has made it his mission to provide that rescue, becoming a science avatar for postwar baby boomers. Death is a bummer. Life is good. Change the paradigm. Hell no, we won't go—ever! He came to Las Vegas because he didn't think life extensionists, cryonicists, hormone mainliners were kooks. He understood them. They wanted rescue, that's all, and until a messiah emerged from the fog of science, they would see what they could do on their own. They were self-starters. Sometimes, perhaps, they were a little misguided. West did not put much stock in the armamentarium of supplements, for example, but as long as science had the same queasy feeling Harley had about addressing the desires of folks who talked openly—and gleefully—about extending the human life span, what could you expect them to do, just accept death? West didn't, and he didn't see why anybody else should either.

At the time West began his quest, hardly anybody was studying aging. Plenty were working in the field of geriatrics. Florence

Mahoney's National Institute on Aging had veered from her own vision and begun handing out grants to investigate the diseases of old people, and how to make the old feel better about being old, but very few were researching how human beings became old, aging as a progressive disease. A scientist named Alex Comfort (better known as the author of *Joy of Sex*) had written a little, and a UCLA biologist named Roy Walford had experimented with mice which he fed very-low-calorie diets, an old technique that curiously extended the normal life span of lab animals and had caught on with some people who went on their own calorie-restricted diets. But when West started looking for a lab where he could start his mission, the pickings were pretty sparse.

West knew little about the science of aging except that a genetic disease called progeria was regarded by some biologists as a surrogate model for aging. In progeria, the joints become frozen with arthritis, the skin wrinkles, the hair grays and falls out, the bones become brittle, and the patients often die of organ failure when other people their age are getting driver's licenses. West reasoned that if progeria really was a twisted version of normal aging, then aging had a genetic basis and some intervention involving genes might be used to slow it down or even stop it.

When he looked for scientists working with aging and genes, he found Sam Goldstein at the University of Arkansas in Little Rock. Goldstein was using gene engineering and molecular biology to look at aging cells, work he had started in Canada in the late 1970s, often collaborating with his student, Calvin Harley. Goldstein reported that a particular genetic sequence he called "inter-ALU" was lost as cells aged. This appeared to be a portal into the aging mechanism and West begged Goldstein to let him come down to Little Rock.

Goldstein accepted West, but when West concluded that inter-ALU was a dead end, the two had a falling out and West moved on to Baylor medical school, where he worked on his Ph.D. in cell biology. Then, while still writing his doctoral dissertation, West enrolled in the University of Texas Southwestern and entered the M.D. program. He was still interested in pursuing theories about cellular aging, so he convinced two biologists working on the problem, Woodring Wright and Jerry Shay, to let him work in their lab while he attended medical school.

At the time, Shay and Wright were becoming increasingly convinced that tiny sections of DNA at the ends of chromosomes, called
telomeres, were the fraying split ends of the genetic program. Every
time a cell divided, the telomeres frayed a little more, becoming
shorter and shorter until, finally, the cell could no longer divide.
Once that happened, the cell either died, or it senesced, became old
and tired, not doing much of anything besides sitting there emitting
funky, misshapen proteins that wrecked the environment around it.
As more and more cells suffered the heartbreak of split ends, bigger
and bigger chunks of tissues looked frizzled, aged, worn-out.

West dismissed the idea. He was enthralled by another theory of
cellular aging until Shay convinced him to visit Harley at McMaster.

In 1985, a colleague of Harley's named Carol Greider used a tiny
freshwater protozoan called a tetrahymena to study cell division. She
discovered that an enzyme called telomerase worked like hair condi-
tioner to smooth out split ends and let them grow longer. It actually
added to the length of telomeres. So the cells didn't die and they
didn't grow old.

It turned out that some human cells use telomerase, too, and
when they do, they don't die either. Scientists studied a collection of
cells called HeLa, named for Henrietta Lacks, a Baltimore woman
who died of cervical cancer at Johns Hopkins in the 1950s. Some of
her cancer cells were preserved and propagated, the first human
cells ever turned into a "cell line," a batch of copies that could be
sent to other scientists for study. This was possible because HeLa
cells are immortal. They just keep living and dividing because, as the
scientists discovered, the cells make telomerase to keep the telomeres long and luxurious. The only other human cells that make
telomerase and can be considered immortal, West discovered in his
discussions with Harley, are the germ cells, the cells that make eggs
and sperm.

The telomere idea of aging insinuated itself into West's mélange of
creation myths. The germline is immortal. Eggs and sperm have no
dead ancestors all the way back to the origin of life on earth. So death
was not automatically a part of life, West reckoned. Death arrived
when cells began to organize themselves by function. For the earliest
living things, there was no death. But then, as evolution rolled along,
germ cells became segregated just as Schrodinger had said. They

stayed in a protective pocket, happily enjoying their immortality, while the other cells became specialized drones made for eating, digesting, carrying on specific functions of life. They could wear out. Germ cells didn't. "This is a very old idea in religion," West explained. "Especially the ancient Mediterranean religions based their hope for immortality on an immortal renewal of life. Even Easter is a reflection of that." Life escapes the germline, but maybe if life could revisit the germline, life could be perpetually renewed.

West found fertile ground for this theory in his adopted home of Texas. Historically libertarian in outlook, tolerant of eccentrics, Texas probably had a higher concentration of life extensionists than any other place with the possible exception of California. West cultivated friendships within the life extension community around Houston, getting to know people like NASA engineer David L. Brown, now a California businessman, and Don Yarborough. Credited by Saul Kent with being "among the other people who tried to promote cryonics in the early years," Yarborough was famous in Texas for having just missed being elected governor when he lost by a thread in the 1962 Democratic primary to John Connally.

A Houston oil geologist named Miller Quarles gave West the boost he needed and, in the process, helped give the world its first biotech company self-consciously devoted to increasing the human life span.

Quarles, a native of southern California, spent most of the Great Depression at Cal Tech in Pasadena, taking geophysics, mostly, but also attending courses taught by Linus Pauling and Thomas Hunt Morgan, the man who created genetic mutations in fruit flies through selective breeding. The idea of genes as units of control fascinated Quarles, and though he spent the next fifty years hunting for oil all over the world, he never forgot these early lessons or Cal Tech. To this day Quarles, now a wiry eighty-nine years old, wears his Cal Tech belt buckle with great pride.

Just about the time West was entering his final year of medical school, in 1989, Quarles started the Cure Old Age Disease Society (COADS). He had two motivations. He was old, for one thing, and had a lust for life—and women—that seemed incongruent with his advancing age. Second, his daughter, a biochemist, urged him not to spend his retirement years doing something frivolous, like writing the book on flirtation he had always wanted to write. "What do you want me to do?" he asked her, "find the fountain of youth?" Well, actually, finding the fountain of youth seemed much more meaningful and when Quarles thought back on his days at Cal Tech he figured that the idea was not nearly as outlandish as it sounded.

Quarles's dentist joined COADS and talked it up among his patients, one of whom happened to know of the work Mike West was doing with cellular aging at Baylor. Quarles called West and the two agreed to meet at an upcoming Alcor cryonics meeting. Quarles told West, and another scientist named Don Kleinsek, that he would be willing to finance a company devoted to life extension. Both men turned him down, with West insisting that the government would never allow any grants to flow to a company that had longevity as its main research agenda. When West returned from McMaster fired with enthusiasm for telomeres, however, he had changed his mind, reasoning that the government would be willing to help because telomere biology might provide insight into cancer. West called Quarles and said he was ready to start a company. Quarles gave him $50,000 in initial financing. West took a leave from his last year of medical school, changed the corporate charter of a little truck leasing company he had started during graduate school, and turned it into a biotech. When he named the company, he referenced the Bible—John 3:3: "'Verily I say unto thee, except a man be born again, he cannot see the kingdom of God.' Nicodemus saith unto him, 'How can a man be born when he is old? Can he enter the second time into his mother's womb and be born?'" This was exactly what West was going to try to do, use the lessons of the germ cells and their always-on telomerase to send cells back in time, back to the "womb." In Greek, the "old man" Nicodemus referred to was called "geron."

West then went looking for more money from venture capital and found Alan Walton. Walton's Oxford Biosciences set West up at an influential venture gathering organized by Ernst and Young and Walton's company. "He just rolled out the red carpet for me. I cannot remember much of what he said during his introduction, but it was, like, 'Here is Mike West talking about the future of biotechnology, the greatest opportunity in biotech, novel interventions in aging.' It was just a wonderful intro! I gave a pretty good talk, one of my better days." Though biomania had lost some of its early blush, and there was not much interest in creating more Genentechs, Amgens, or Biogens, investors were still looking for new ideas and the phrase "novel interventions in aging" made a killer "story." His talk so impressed some partners from Kleiner-Perkins they surrounded West as he walked out the door, insisting he let Kleiner-Perkins lead the first full round of investment. West dropped out of medical school, hired Harley and another scientist, and set up shop in Dallas. He soon moved to Silicon Valley, closer to Kleiner-Perkins and the science talent of the San Francisco Bay Area.

As Wally Steinberg had proven, when it comes to making money in venture capital, it doesn't really matter if the company you finance can really do what you say it can do. All that matters is that you can make a bold statement and then get somebody to believe it. Stopping aging sounded outrageous. But "interventions in aging" was compelling and the telomere-cancer link gave Geron a near-term goal to shoot for. So West, a newly minted Ph.D., a man with few publications to his credit, found himself being handed millions of dollars and the reins of a biotech company.

Unfortunately for West, those reins were also strings. He was no longer a footloose postdoc, nor a med student who was free to think and say anything he wanted about Osiris and Isis and returning life to the germline and curing aging. He was now the founder of a real company that had used other people's real money and other than Quarles and West's life extension buddies back in Texas and within the cryonics community, few wanted to hear anything about curing death—especially the board of Geron, which controlled much of the company's voting stock. It wanted to see a business plan that contained something other than making people live to be 300 years old. Cancer, of course, was the obvious place to start. Testing for the presence of telomerase could be an early diagnostic for cancer, so Geron worked on developing a test. Maybe preventing cancer cells from making telomerase would stop them from dividing out of control, keep cancer from spreading, so Geron explored how to shut down telomerase through drugs or gene therapy. These were reasonable aspirations.

Tension between what West wanted and what the board of directors wanted was built into Geron. Just about the time he appeared at A4M in Las Vegas, that tension was increasing, because West, who has become well known within biotech for developing intense infatuations with one idea after another, had a new idea. One busy day he had gone for a walk along the shore of San Francisco Bay to clear his head and by the time he returned to the office, he had decided on a course that would not only lead to him being booted out of Geron, but would set biotechnology on a collision course with the president of the United States and compel Bill Haseltine and others to argue that the biotech rapture had finally arrived.

IN 1994, WHEN WEST TOOK HIS WALK along the shore of San Francisco Bay, he noodled over how to build a bridge like the one Watson had tried to build at Harvard and Haseltine was in the process of building at Human Genome Sciences. "I started thinking about this dichotomy, this diagram: immortal germline, mortal somatic [body] cells," West said. He reckoned that there had to be a way to traverse the gap between the two, to go backward into evolution and reunite the body's cells with the everlasting life they had lost when the germ cells became segregated in the earliest living creatures. Maybe, he thought, you could isolate and grow the cells of development, cells that would presumably have the telomerase gene switched "on," but would not be cancerous.

This was not really a new concept. The early developmental biologists had watched animal embryos in dishes, and, by 1994, the world already had sixteen years of experience making human embryos in dishes, thanks to the successful IVF experiments of Robert Edwards and Patrick Steptoe. Biology knew that the earliest cells of development were extremely powerful, but it did not have a way to experiment on them.

The first surrogates for these cells came from odd cancers found in the testicles of mice called teratocarcinomas. A cancer of the cells that give rise to sperm, a teratocarcinoma makes a tumor called a teratoma. When work with mouse teratomas began in 1959, it was discovered that they make embryos from Bizarro World. Scientists dissected the teratomas and found whole teeth, bits of bone, hair, tiny blood vessels, and not simply in random bits, either, but in sad attempts at building real systems. A teratoma might include cartilage attached to muscle, say. In other words, when the germ cell turned cancerous, it started to try to make a baby. But because the signals were scrambled, it wound up creating oddball tumors instead.

Such human cancers had the same strange property, the desire to make babies. Cells from mouse and human teratomas, called embryonic carcinoma, or EC cells, were isolated in the early 1970s and used to make cell lines to study how a mouse became mouse and a man became man—the point, after all, of developmental biology. How does a creature, a mouse, a man, a fish, for that matter, get made?

The real answer is so fantastic that it is one of the few realms of biology that still inspires awe among scientists. We start as a single fertilized cell called a zygote, the result of the fusion between sperm and egg. Everything that we are, every cell, every tissue, every nerve fiber in our brains, comes from this single cell in an intricately choreographed ballet of self-assembly, as if a bolt were placed on a garage floor and grew into a Ferrari. When the zygote divides, it becomes an embryo and one of the embryo's first jobs is to turn itself into a semihollow sphere. Some cells migrate to the edge of the sphere and will eventually build a placenta. Meanwhile, another group of cells inside the sphere, called the inner cell mass, will make a baby. These are embryonic stem cells.

When teratomas were being studied, science thought they could be the Rosetta Stone for decoding how this self-assembly worked. Odd, yes, but one way of looking at pregnancy was as a kind of controlled cancer and the thing about EC cells was that, as corrupted as they were, they were obviously capable of building most, if not all, of the various tissues in the body. So throughout the 1970s, biologists figured they could use EC cells as a model to view the process of development and answer the question of how cells from the inner cell mass decide what to become. Why does one cell go down a road that leads to making a liver, while another cell heads toward bone? This process, called differentiation, was hard to figure out in real embryos because the changes occur on minute scales moment to moment by communication among dozens of cells, so EC cells became surrogates. The developmental biologists hoped that they could use some of the techniques of molecular biology to determine which genes turned on or off and in what order to direct development. They wanted to capture "factors"—the term biology made up for unknown "stuff"—that were telling cells what to become. How a single cell gave rise to life was one of the great unanswered questions of biology.

"The big question we had in Watson's lab, the one we all thought we were working on," Haseltine recalled, "was differentiation and how a body is constructed through the programmed expression of genes." But like much of the rest of developmental biology, this took a back seat to the glamour sport of *E. coli* and phages and the quest for biological control over these simple systems.

The developmental biologists, though, did not forget. Almost as soon as they had isolated the first sustainable human EC cell lines, they wanted to create the real McCoy, colonies of human embryonic stem cells. After 1978, and the birth of the first IVF baby, there was no particular reason why you could not pull ES cells out of a human embryo and make them grow like EC cells were made to grow. There was not much ethical discussion about this, either. IVF made more embryos than it used, embryos that would wind up destroyed, so cracking them open for the ES cells inside did not raise many eyebrows. Logistical problems, not ethics, foiled the plans of Peter Andrews, an English scientist who was then working at the Wistar Institute and who had been among the first to isolate human EC cells. He could never seem to get hold of any embryos, even though he started looking as early as 1981.

That was the year that the London lab of Martin Evans announced that they had isolated and grown mouse embryonic stem cells. (Like almost every big moment in biology, the "discovery" of stem cells was the result of work done by dozens of people over many years and there are competing claimants, including Robert Edwards, who told me he was first to isolate stem cells from rabbits in the 1960s. An American, Gail Martin, a former member of Evans's lab, also published her own isolation of mouse stem cells at about the same time Evans did, but Martin Evans is the man who generally gets the credit and, as he told me when I visited him in Cardiff, Wales, where he now directs the university's biology program: "I certainly did discover stem cells!")

Evans had searched for mouse ES cells because he wanted to study differentiation in the ultimate lab animal. But the cells were almost immediately hijacked by molecular biology. It turned out that embryonic stem cells were a good tool for making new kinds of mice and then watching how genes worked in real live mammals. Two other scientists, Mario Capecchi and Oliver Smithies, who both shared the 2001 Lasker Award with Evans for this work, genetically engineered the stem cells, placed these cells back into mouse embryos, and got custom-made mice. The idea of patenting the process for making the ES cells never entered Evans's mind, so the cells circulated freely among many scientists who, for the first time, could build mice pretty much on demand to act as models of human disease. Need a diabetic mouse? "Knock in" the gene that causes diabetes. Need a hairless mouse? "Knock out" the gene for hair. Need a mouse to study the immune system? Make a mouse without one.

"All our mind-sets turned to using these as vectors to make transgenics," Evans said with a note of regret in his voice.

Mainstream scientists often made their reputations based on work with lab animals, not people and not cells from people, which meant that biologists often knew far more about mice than their fellow man. This was a way of thinking that Haseltine, for one, objected to and why he maneuvered to work on human disease. But for many biologists, the mouse was king, and now that mouse ES cells were available, and mouse EC cell lines existed, few saw the need to bother making human ES cell lines. As Evans said, "There was no impetus to do it . . . We said, 'There is no point.'"

This attitude was a carryover from the decades before Evans's work, when developmental biology was pursued simply for the sake of solving the riddle of creation. Developmental biology, like all biology, had been a gentleman's pursuit, mainly, meant to scratch the itch of curiosity. Even after World War II, when molecular biology grew and people won Nobel Prizes for their work, the developmentalists had no real interest in or expectation that what they were studying in frogs or mice or chickens could ever have any application to people.

Peter Andrews at Wistar had trouble even getting money to work on the EC cells. There was no way he could find money for human ES cells because grantors thought such work would be irrelevant to humans. What would you do with them? Answering questions about development might be interesting, but scientists were not about to start making designer people the way they made designer mice, so why spend the money? "It was four years from when I first went to Wistar to when I first got an NIH grant to work on [human] EC cells. The common criticism you got was 'There are mouse EC cells and mouse ES cells, why bother with human [cells]?' . . . So you have a whole period of about twenty years during which few people actually did things on ES cells or EC cells for their own sake."

With much of biology diverted from embryos and stem cells, the field was left to rebels and outcasts. Government grant makers weren't about to fund research into human ES cells, and neither were universities. The bailout option for scientists who wanted money to pursue edgy work, industry, wanted no part of it, either. There was only one place, the world of agriculture, where ES cells, eggs, and embryos were not just interesting curiosities of basic science, but valuable tools meant for practical use.

Benignly ignored as country cousins, ag scientists performed much of their work while wearing gum boots, an arm up a cow's nether regions, a world away from the high falutin' culture of the molecular biologists. From the time of John Hammond, these scientists never lost the focus on stem cells, eggs, sperm, and the miracle of development because they could see all kinds of practical applications for those mysterious processes—making a better cow, a better sheep, a better horse. And because they did come from a place where nobody was looking, a community that rarely made headlines, they were able to engage in some of the most radical biology of the century, sometimes just for fun.

Hammond's influence spread not only through his experiments but also through the many students he trained at his lab in Cambridge, England, a line of descent that eventually produced a young graduate student named Ian Wilmut. In 1966, during Wilmut's final summer vacation from what was until then a lackluster college career studying agriculture (when Wilmut tried plowing he was unable to make a straight line), he won a scholarship to work as an intern in Cambridge. At the time, the lab was studying embryos and the preservation of cells by freezing. Wilmut came away fascinated and inspired. He took his undergraduate degree and then returned to Cambridge for his Ph.D. His dissertation was on the freezing of boar semen. He went on to become the first to produce a calf from a frozen embryo in 1973, five years before human IVF succeeded.

Immediately afterward, Wilmut left Cambridge to work at the Roslin Institute, an agriculture research station just outside Edinburgh.

Though far from the limelight of big-name academic science, agricultural researchers like Wilmut were at least able to experience a rarity in biological studies: immediate gratification. They did not have to spend twenty-five years working in some small niche of biology in hopes serendipity would lead to an important discovery. They could see real results quickly. Crossing every molecular T, dotting every protein I, mattered less than whether or not you got a good result somebody could use. And since they worked with farm animals, nobody was going to squawk if they mixed cells from sheep and goat embryos just to see what they'd get (a "geep," as it turned out). They could do pretty much any damn thing they wanted.

Because the subjects were not people, not even lab mice, materials were often cheap by comparison to those used by their academic cousins. Large animal scientists did not have to worry about wasting precious cow eggs, for example, because their animals marched into a slaughterhouse, where ovaries could be scooped up by the bucketload, not tiny ovaries with tetchy, fragile eggs from highly inbred strains of lab mice either, but big, fat cow and pig ovaries. So they were able to push their work into frontiers the more famous university scientists could not or would not. As a result, the ag guys were among the first to realize just how powerful the egg could be and the possible practical uses of embryonic cells.

Most of the world was unaware of what they were doing. These scientists almost never talked with the big-name academics, some of whom dismissed people like Wilmut as merely "vets." Unlike Haseltine, whose transformative moments came courtesy of a string of Nobel Prize winners at Harvard and MIT, Wilmut's life-changing scholarship to that Cambridge lab arrived courtesy of the Pig Industry Development Authority.

As that name implies, the work of large animal reproduction specialists was often aimed at making new varieties of pigs, cows, goats, to produce more meat, wool, milk, or, in later years, to create transgenics, the way the molecular biologists had done with lab mice so that a cow, say, or a sheep, could produce a human protein in its milk that could be extracted and used as a drug for people. In the 1980s, after Evans's isolation of the mouse ES cells, the ag world, including

Wilmut, began trying to work with ES cells from their animals. If they could grow them, they could engineer the desired traits and then put ES cells into early embryos, just as the mouse people had done. But ES cells proved elusive. Nobody could get them to grow.

In a kind of shoot-the-moon effort, some tried a process called nuclear transfer, moving the nucleus of a cell (or sometimes just a whole cell) into an egg. In the 1930s, Hans Spemann, the discoverer of "organizers," factors that influenced development, had speculated about such things, suggesting a "fantastical experiment" to test whether a cell from a fully formed adult, say, a skin cell, could be transplanted into an egg to create a whole new animal, a clone. John Gurdon would carry out such an experiment with his frogs in 1962, but nobody had done it with a mammal.

Now though, frustrated by their inability to culture farm animal ES cells, the ag world resurrected the idea and quickly succeeded in producing clones by inserting cells from early animal embryos into animal eggs. In 1989, Wilmut led the team that produced the Roslin Institute's first sheep clones this way, proving that at least the earliest cells of development were able to make a whole creature. Then, in 1991, Wilmut was joined at Roslin by a young postdoc named Keith Campbell. Campbell, intrigued by cloning ever since he studied Gurdon's frog experiments, came to Roslin hoping to follow up on the master's work. He wanted to make a clone with cells from grown-up animals.

While the large animal scientists were speeding ahead, the few mainstream scientists who wanted to continue to press the hunt for human ES cells were forced to look outside the usual channels of the NIH and universities. That's where Mike West lived. As a product of the life extension movement, West was not simply unafraid of working in the human realm, he was all about humans. To West, humans, not mice, not bacteria, not yeast or fruit flies or worms, were the point of biology.

HARLEY HAD LONG USED the aphorism that "babies are born young," a statement that seems banally obvious but is actually a paradox. How is it that very old sperm and very old eggs from older people can do the mating dance of life and wind up making very young babies? Something must happen to those cells to make them young again. West thought the first cells from that union "might be naturally telomerase positive. They might be young."

In the November 9, 1994 issue of the journal *Human Reproduction*, a Singaporean named Arrif Bongso announced that he had done just what West was hoping could be done. Bongso, using twenty-one donated embryos from nine IVF patients, managed to grow and then isolate some lumps of inner cell mass. He then separated the cells and grew them in a dish. He reported that they sure looked the way genuine human embryonic stem cells ought to look. Unfortunately, they did not last long. Bongso's cells died after two "passages," or divisions. But his work, published in a journal read by many IVF doctors, alerted others that in Singapore there was a scientist who was not afraid of working with human embryos.

Since Bongso's human cells died, West thought it might be a good idea to do some preliminary studies on mouse ES cells, then work up to human cells. So West went to visit Roger Pedersen at the University of California at San Francisco, hoping Pedersen could provide his mouse cells.

During their meeting Pedersen happened to mention that he was also the director of UCSF's human IVF clinic. "I said, 'Oh?'," West recalled, his eyes lighting up the way they must have lit up that day. "I said, 'Roger, did you ever think about trying to make a human ES cell?' It was like I had just said his mother was from Mars. He just completely turned off. I thought, holy cow! maybe he's offended. So I changed the subject."

Pedersen's reaction was predictable. West had lifted a curtain on a taboo subject. The use of human fetal material was already mired in American abortion politics, a debate that led to gross-out speechifying about cannibalizing babies to save our own skins. Pedersen, who sometimes received NIH grants, knew that the idea of using human embryos as a source for cells in experiments would be just as explosive. Already, another scientist named John Gearhart at Johns Hopkins was experimenting on cells taken from aborted fetuses to see if he could isolate a version of stem cells. Gearhart was getting death threats for his trouble. In this atmosphere, the science tribe had developed ways to discuss such things, and it wasn't by bringing up the topic with people you've just met while sitting in an office the government helps support with grant money. Instead, such discussions took the form of oral samizdat, whispers among colleagues over drinks in hotel bars during off hours at scientific meetings.

In contrast, the ag guys were busy trying all kinds of crazy ideas. By the early 1990s, some had started performing the experiment Campbell wanted to try, taking adult cells from animals, putting them into eggs, and seeing if they could make a clone. At least one lab got pregnancies in cattle this way, though no calves were born. Then, just for grins, some swabbed the inside of their own cheeks to get a cell sample. They cultured these cells, shoved them into cow eggs, and stood back to see what would happen. Sometimes, the cow eggs turned the human cells into embryos. "It was really fun," one would-be cloner recalled.

The parallel tribal grapevine of the mainstream, human-centered scientists did transmit the occasional buzz about human ES cells. Everybody knew about Bongso's work, of course, and though Gearhart was keeping a low profile, insiders knew what he doing. James Thomson, a former colleague of Peter Andrews's from Wistar, reported to colleagues that his new lab at the University of Wisconsin was close to culturing ES cells from monkeys. But most of this work was being done on very tight budgets.

Still, a critical mass was beginning to develop around the idea of culturing human ES cells. Eventually, biologists in the field knew that if they did not start working now, somebody else would be first. This was an important consideration because other technologies, like tissue engineering, had developed over the previous decade. So human ES cells were no longer just an academic pursuit. If somebody got them, the cells might be used to literally grow new organs for humans. Maybe they could be used to treat diseases in which cells go bad, brain diseases like Parkinson's or Alzheimer's. Maybe they could tell biologists something more about cancer since cancer starts with a cell that, for some reason, turns itself into a stampeding version of a stem cell.

By 1996, Pedersen had a change of heart. "He calls me up," West remembered, "and says, 'Mike? Remember you asked me about human ES cells? You said you'd be willing to sponsor some work to get them done? I am open to that. I've changed my mind.'"

In quick succession, West signed deals for Geron to finance Pedersen, Thomson, and Gearhart. But the Geron board went along grudgingly. "If this had been Wally Steinberg, he would have just said, 'Go do it,' and put up the money," West said, but "there was no interest in Geron." That attitude was certainly understandable. The company's telomerase program had not yet found the telomerase gene, and the venture capitalists were getting antsy for Geron to go public so they could take their profits and bail out. Now West was talking about a whole other program for stem cells? The board was also increasingly restless about West's immortality talk. The message to the outside world was supposed to be about cancer and maybe "the diseases of aging," but nothing that sounded as flaky as the language coming out of the alternative medicine crowd at gatherings like A4M. So while the board ultimately acquiesced, it was none too pleased.

In August 1995, Keith Campbell, Wilmut, and the rest of their Roslin team announced the birth of two lambs, Meagan and Morag. These lambs were clones. They had come from a line of fetal cells grown in a dish. A biotech company, PPL Therapeutics, located next door to the Roslin Institute, had been trying to develop ways of making sheep and cows produce human proteins in their milk and the company immediately realized that the Roslin technique that produced the twin lambs might be a powerful way to make animals more efficient drug makers. So PPL gave Roslin money and samples of mammary gland cells PPL had been studying from an adult female sheep. Campbell was convinced that adult cells could make a clone and, thanks to the biotech's money and cells, Roslin was able to try. It took a while, but Campbell's idea worked. A lamb named Dolly was born on July 5, 1996, the first mammal ever created as a clone from the cell of a full-grown adult.

Roslin kept her birth quiet. The institute knew the *Nature* paper they had prepared would become as famous as the Crick and Watson paper describing the double helix in 1953 and they did not want to mute the intrascience impact by letting the lay press splash the news before the paper could run. They almost made it. The *Nature* paper ran on February 27, 1997, but the London *Observer* headlined it first, four days early. From that moment on, Dolly became the most famous animal in the world.

Many members of the agriculture science clan were not shocked, really, that the Dolly experiment had succeeded, because they had been messing with adult cells, sometimes their own, and animal eggs for a while. But the big-name scientists outside agricul-ture were amazed, even Gurdon.

To West, Dolly was a godsend, the answer to a problem he was desperate to solve. In 1996, while Dolly was secretly in utero, Geron mounted an investor road show to drum up interest in its public offering. The smarter stock analysts questioned Geron's financing of the stem cell search just like the Geron board had. Even if Pedersen or Thomson or Gearhart succeeded in growing human ES cells, what good were they other than as an interesting science experi-ment? New organ factories and cell therapies sounded good, but there would be serious obstacles. When mouse ES cells were injected into mice, they made teratomas, cancer, so you could not just shoot stem cells into people. Even if you could, the patients' immune system might reject them just like they rejected organ trans-plants. You'd still need drugs to prevent that rejection. So how did Geron expect to make money from human stem cells?

When the news of Dolly arrived, West realized that Geron could make ES cells compatible for patients by taking a patient's own cells, putting them into an egg, making an embryo, and then taking ES cells out. The ES cells would be the patient's own. No rejection.

In July 1997, John Gearhart announced at a science meeting that he had cultured human embryonic germ cells, the cells that will eventually produce sperm and eggs. These cells had most of the fea-tures of ES cells.

Meanwhile, Thomson was using a system almost exactly like the one Evans had used to derive mouse cells and was trying to tease human cells out of embryos supplied by a scientist at a tiny hospital in Israel. Thomson finally managed to get some cells out and grow them, but he couldn't be sure these were real human ES cells. So he sent some to Peter Andrews, hoping that when Andrews compared them with his EC cells, all the same telltale signs of "totipotency," the power to become any cell in the body, would be there.

They were. On November 5, 1998, almost exactly four years after Bongso's original paper, and West's decision to fund an ES cell search, the University of Wisconsin office of communications released the news of Thomson's achievement. "The dream of one day being able to grow in the laboratory an unlimited amount of human tissues for transplantation is one step closer to reality," it began. The release went on to list any number of uses for ES cells, from treatment of Parkinson's disease and diabetes, to heart disease and cancer.

There was a one-sentence caveat. "Such clinical applications are years—perhaps more than a decade—away."

The next day, Geron's stock nearly doubled. On November 4, volume was only 241,000. On November 5 and 6, over 45 million shares changed hands.

A month later, Haseltine told the *New York Times* that Human Genome Sciences' inventory of "the 14,000 signals and receptors . . . would be expected to include those that are involved in embryonic development and hence, once identified, would supply most of the factors needed to make human embryonic stem cells differentiate into any desired type . . . Haseltine said he was discussing a collaboration with Geron."

But West was no longer at Geron to celebrate. The friction between the Geron board and West finally produced too much heat. Eight months before the stem cell announcement, West was fired, nudged, or left voluntarily. Just which scenario is true is still disputed inside biotech, with West insisting he sold his stock and walked out because Geron had become too diverted from his antiaging mission. The Geron press release announcing his departure made nice, as all such press releases do, offering up a quote from the company CEO crediting West as "a source of provocative thinking."

West was out in the cold now, with no place to test his idea of using cloning to make compatible stem cells, of returning life to the germline. Once again, fate took a hand.

Several months after he and Geron split, West saw a presentation during a scientific conference in Australia about an experiment in which ag scientists took cells from a forty-day-old cow fetus, grew them in a culture dish until they were old and had little life left, then put them in an egg, implanted the resulting embryo into a cow, grew another fetus to forty days, took cells from the fetus, and grew those cells up in a culture dish. The cells divided just as many times as the first batch. They had literally been renewed. Not only were they compatible, they were young. The same genetic program had its life extended just like it did in Robert Heinlein's Howard Rejuvenation Clinics.

"I turned to my wife and said, 'Oh my God! That's the cellular time machine! You do not need telomerase!' I had left all my telomerase intellectual property back at Geron, my life's work, but there, plopped into my hands, was the idea that nuclear transfer could rebuild the life span of cells." Almost immediately, West began talking to the leadership of the tiny agricultural cloning company that had done the work, Advanced Cell Technology. Within months, West was the new CEO. He moved to the company's location, Worcester, Massachusetts, right next door to where Haseltine's Cambridge Bioscience had sat.

In May 1999, Geron bought Roslin Biomed, the spinout the institute had set up to manage the cloning patents used for Megan and Morag and for Dolly. Geron, it turned out, thought West's idea for making compatible stem cells was a pretty good one.

But just when man appeared ready to take the biological reins, to finally be on the verge of fulfilling the prophecy of Haldane, some said the whole thing was a big mistake, that the ongoing human project to control biology would lead to a slow-motion holocaust. There had always been opposition to fiddling with nature. At the beginning of the Renaissance, clerics argued the dissection of bodies was a sacrilege. *Frankenstein* was written as an argument for the supremacy of the sublime in nature over the possibility that new electricty experiments might defy nature by "reanimating" tissue. John Hammond's artificial insemination was held at bay by the Church of England. And, of course, there was *Brave New World* after Haldane.

Until stem cells, Dolly, gene engineering, and the move toward turning genetic information into a commodity, these arguments were mostly academic. Electricity did not, in fact, reanimate dead tissue. Now though, science fiction did not seem so fictional. It seemed possible that human beings might finally escape natural evolution and take control. As a result, the debates over these developments took on an urgency and passion they had never known. A battle was about to begin for the hearts and minds of the public, most of whom had never given biology and biotechnology a second thought.

7   Bring On the Inquisition

THE RAPTURE MIKE WEST HOPED to deliver was now more real than ever to the bioutopians. After a century of gene research, new genomics companies like Haseltine's Human Genome Sciences were finally promising large-scale control over the molecular drivers of the human body. By 1995, HGS claimed to have used the data from TIGR and from its own scientists to isolate and characterize 95 percent of all human genes. By 1998, Haseltine said the company's scientists had identified 14,387 genes that encoded the instructions for making proteins which, in turn, passed orders on to cells to tell them what to do. The earliest cells of human development, the cells that made everything, had been isolated and cultured. Dolly proved just how powerful those cells were, that they could be made youthful over and over again, and that they were plastic, able to be shaped and molded according to human desire.

What's more, there was an undercurrent among scientists that the potential long-range future envisioned by the bioutopians not only could come true, but might be desirable. After a century of hype, phoney science, and sham cure-alls, biologists were not prepared to state that case just yet—they still held the bioutopians at arm's length. But the feeling was growing.

All this had seemed so improbable just a few years before that people who paid attention to such things suffered from a sudden vertigo. Before 1992, hunting for human genes was a long, slow slog and the genome project was sold as a long-term human endeavor, like the space program. But then people like Wallace Steinberg, Craig Venter, and William Haseltine turned genes into patentable commodities. Michael West showed up out of the blue all the way from a truck dealership in Niles, Michigan, a creationist bible college, and the immortality fringe, as likely a story as the dwarf, Eddie Gaedel, coming off the bench to hit for the St. Louis Browns. Though Florence Mahoney had tried, mainstream science had considered the how-to of aging to be the roach motel of science careers. People who went in never seemed to come back out, so only a handful of biologists even worked in the field. Now, thanks to the deus ex machina of venture capital, and the intercession of a few wealthy immortalists, there was talk of companies being formed to combat it with real science. West's dream of creating an antiaging business had been truncated and Steinberg died before he could start one, but the idea of vast life extension was brewing. Throw in human embryonic stem cells and Dolly, and it was no wonder that biotech was increasingly being seen as a new religion, the answer to the big questions surrounding death, disease, and human transcendence.

But as with every religion that blooms anew, there were those who condemned the belief in biotech as the worst of heresies. The increasing likelihood of human control over human evolution was too much for the "bio-Luddites."

At least since the dawn of the Christian era there have been bio-Luddites. But the gene splicing era rekindled opposition to experimentation with the stuff of life. When word of recombinant DNA experiments in Paul Berg's Stanford lab leaked out, some scientists were concerned that Berg risked creating a kind of superbacteria that could unleash a plague. In one of the most important milestones in the history of biology, researchers gathered in 1973 at a California conference center called Asilomar to decide on safety procedures for continued gene splicing experiments, and to assuage the public fear that science had gotten out of control. Today, the biologists involved look upon "Asilomar," as it is known, as a shining example of the responsible exercise of scientific caution.

One man who pled for a go-slow approach was physician and philosopher Leon Kass. Kass was already well known in science for his early opposition to IVF, arguing that it would amount to the manufacturing of people and that it could not be considered a therapy because it did not cure infertility, a criticism its inventors, Robert Edwards and Patrick Steptoe, dismissed out of hand by citing insulin and eyeglasses as examples of noncures that were, nonetheless, valuable treatments. In October 1971, at a symposium in Washington, D.C., entitled "Fabricated Babies: The Ethics of the New Technology in Beginning Life," Kass was part of a discussion panel that included Watson in his role as oracle of all things biology, Princeton theologian Paul Ramsey, American IVF pioneer Howard Jones, a lawyer, and an embryologist.

Ramsey launched an attack on Edwards "as if from some nineteenth century pulpit," Edwards recalled in his memoir, *A Matter of Life*. Kass followed, condemning Edwards's work out of fear that babies would be born horribly deformed. Edwards countered that Ramsey's objections were right out the playbook used by those who had condemned Darwin and defiantly told Kass and the other objectors that his IVF experiments would continue. "This plea for our experiments to stop is ultra-conservative and unacceptable."

Kass has been among the most conservative and vocal opponents of biotech ever since that symposium, denouncing everything from IVF to the Supreme Court patent case that jump-started the biotech industry in the first place. After IVF went on to produce thousands of healthy babies on the heels of the first Edwards–Steptoe success in 1978, and after biotech produced valuable drugs like recombinant insulin, Kass had come to be seen as something of an anachronistic zealot in the science and bioethics establishments.

Still, there were other bio-Luddites who arrived at some of the same conclusions Kass had, but through different reasoning. Some resurrected Mary Shelley's argument about the sanctity and power of nature. Others used a social equity argument. Just the rich, they said, could afford the new technology. What all the bio-Luddites had in common, though, was a fear that the precise future dreamed of by the bioutopians might actually come true. As the bioutopians adopted writers like Heinlein, the bio-Luddites reached back through the decades and made Aldous Huxley's *Brave New World* their bible, chanting the title like a mantra to tag any number of technologies as inherently evil. (Huxley's later work, like his endorsement of reaching for enlightenment via hallucinogenic drugs, the *moksha*-medicine depicted in the novel *Island*, was ignored, as were his later statements of approval for biology and science in general. *Island*, in fact, contains praise for modern biology from Huxley, who was, after all, T. H. Huxley's grandson.)

The reality of cloning and stem cells pulled bio-Luddites like Kass in from the margins and galvanized a strange coalition between conservative politicians, Christian evangelists, the Catholic Church, left-wing intellectuals, and green environmentalists, all of whom realized, like the bioutopians, that gene technologies, welded to stem cells and cloning, might finally permit humans to decide their own biological future. With cloning technology it was now possible to genetically engineer a cell with some desired trait, insert that cell into an egg, and get a custom-made creature. That's why it was invented. Stem cells made that prospect even simpler, just like they had for making customized lab mice. Those prospects drove the unlikely alliance. Some of the right-wing intellectual bio-Luddites, like Kass and social theorist and author Francis Fukuyama, held meetings under the auspices of something called The Bioethics Project which was organized by conservative magazine editor William Kristol. This, naturally, made the liberals queasy. "These are the last people I want to be aligned with," Boston University lawyer and bioetchicist George Annas said about his conservative partners in the informal federation. "It's frightening to see those people lined up with me, but you cannot pick your friends or enemies. We have some very strange comrades here, no doubt about it."

Strange, maybe, but the coalition worked. Freaked out by the seeming weirdness of Dolly and cloning, the public was ripe for the message of bio-Luddism proclaimed by Kass.

"Human nature itself lies on the operating table," he warned in his 2001 book *Life, Liberty and the Defense of Dignity*, "ready for alteration, for eugenic and neuropsychic 'enhancement,' for wholesale redesign. In leading laboratories, academic and industrial, new creators are confidently amassing their powers and quietly honing their skills, while on the street their evangelists are zealously prophesying a posthuman future. For anyone who cares about preserving our humanity, the time has come to pay attention." The rhetoric sometimes sounded a lot like Robert Welch's old John Birch Society Blue Book warnings of impending communist doom. Indeed, the only mention of J.B.S. Haldane in Fukuyama's own book-length warning, *Our Post Human Future*, was as "the Communist, J.B.S. Haldane."

No amount of hyperbole was too much if it succeeded in scaring the bejesus out of the public. Kass even equated the fight against the evils of biotechnology to the battle against international terrorism: "the human future rests on our ability to steer a prudent middle course, avoiding the inhuman Osama bin Ladens on the one side and the post human Brave New Worlders on the other." According to Kass, we really were on the verge of transforming ourselves into posthumans. "We are not yet aware of the gravity of our situation," he warned ominously. Already, the process of posthumanism had begun. "The Pill. In vitro fertilization. Bottled embryos. Surrogate wombs. Cloning. Genetic screening. Genetic manipulation. Organ harvesting. Mechanical spare parts. Chimeras. Brain implants. Ritalin for the young, Viagra for the old, Prozac for everyone. And, to leave this vale of tears, a little extra morphine accompanied by Muzak." In the space of two pages, Kass managed to evoke just about every bugaboo of the twentieth century, even the Nazis.

Kass and the other conservatives became the Vince Lombardis of the biotech opposition, insisting that suffering was part of our human nature. Hurt? Tape it up. Happy pills like Prozac were sucking the manliness out of us. So what if you're depressed? That's life. Kid can't pay attention? Well, some kids are like that. Stop being sissies. Take your imperfections like the true Americans you are and die on time. "Bourgeois realism about the limited aims of human striving—health, self-improvement, commerce—may be conducive to a failure of realism about what man is: both the evils he is capable of, and the vulnerability and need for courage that ultimately define him," wrote another Kass ally, Eric Cohen, in Kristol's *The Weekly Standard*. Health and excellence are "gifts"; some people have them and some don't.

But what if you didn't want to die at eighty? What if there really were a way to extend the human life span? It should be banned.

"So then," I asked Fukuyama, "you would argue that the government has a right to tell you that you have to die?"

"Absolutely," he replied.

Liberals, on the other hand, were worried that not everybody would be able to take part in the coming rapture. Therefore, nobody should. "I am not against, in principle, using any technology we can to alleviate suffering or improve quality of life but in terms of purposefully reengineering the human race, we're not ready," Annas said. "The vast majority could not afford it. The others will come to dominate the rest of us, treat us like slaves, or we will become so threatened we will kill them first. That is the prospect of genetic genocide. Either way those are the two most likely outcomes. Either is bad for humanity."

Yikes! The liberals said we faced mass annihilation. The conservatives said we were slipping into biotechnology pods in the garden shed and when we emerged we'd be the walking unhuman.

As for the greens, well, they just accused biologists of wanting to poop in the gene pool. Nature knew best.

Pop culture was receptive to these messages because it had adopted the gene and come to believe in its power not only to dictate life, but *meaning,* too. A century of science fiction, two decades of biomania during the rise of biotechnology, the Human Genome Project, the genomics-derived biotech boom of the late 1990s—and now both fear mongering and predictions of a new utopia—had turned DNA into an icon. The gene became the immutable platinum meter bar locked inside the body and used to measure all things, shorthand for motivations, causes, effects. Magazine editor Tina Brown claimed to have the "entrepreneurial gene." A top female executive explained the difference between successful men who keep working and successful women who quit by saying "maybe [male] DNA says they'd drop dead if they didn't work." Former substance abuser Charlie Sheen said his father and fellow ex-substance abuser, Martin, might feel guilty "for passing on this gene." An English comedic actor was said to have "the clown's rogue gene." "Is there a special gene among intellectuals that lends itself to the embrace of tyranny?" an interviewer asked a book author. "Do you have the shopping gene?" one writer asked another in a magazine article. Books now had DNA. Houses had DNA. "His guardians were within him, propelling him like a tiny human crew within a tall, walking armature of DNA," concluded a John Updike short story.

Artists and writers joined the tide, making the helix part of the artistic vocabulary. Salvador Dali painted *Butterfly Landscape, The Great Masturbator in Surrealist Landscape with DNA* in 1957, then *Galacidalacidesoxiribunucleicacid (Homage to Crick and Watson)* in 1962. But it was not until the late 1990s that it became a fad. Frank More painted *Beacon* in 2001, which portrayed a lighthouse whose beam was woven with DNA. A few artists, like More (and, of course, Natasha Vita-More), saw biology and biotech as an exciting frontier. At least one, Hunter O'Reilly, a geneticist-turned-artist, was given a teaching post at the University of Wisconsin.

But most seemed troubled. In 2000, a New York gallery called Exit Art mounted an entire show around the theme of DNA, genetics, and biotech. "Bradley Rubenstein shares with us some undetectably manipulated photographs of adorable little kids who have something strange about them," wrote *New Yorker* reviewer Peter Schjeldahl. "They look out at us with dog's eyes. . . . Simply to think along some current scientific lines is to beget monsters, to the point where any monstrosity may assume an air of science." Minimalist composer Steve Reich mounted a traveling show called "Three Tales," a video-music performance that includes scenes of the Hindenburg burning, Bikini Atoll being blown up in a white hot nuclear cloud, and Dolly the sheep. In the most popular movie of 2002, the radioactive spider that had turned Peter Parker into Spiderman in the nuclear-fear age of the 1960s became the genetically engineered spider of the bio age.

The bio-Luddites sensed the angst and forged a new weapon based partly on Kass's notion of "The Wisdom of Repugnance," the gut feeling that some things just sounded so yucky we shouldn't do them. Cloning, its obvious sci-fi implications already well ingrained by two decades of fiction and speculation, was the perfect opening to project the concept. The bio-Luddites should exploit cloning fear, Kass wrote, to "get our hands on the wheel of the runaway train now headed for a posthuman world and steer it toward a more dignified human future."

In the 1970s, before any mammal had been cloned, at a time when most biologists thought it would be impossible to clone an adult mammal, Robert Edwards encountered the cloning taboo. At the time, he and Patrick Steptoe were still trying to develop the techniques that eventually led to the birth of Louise Brown, the first IVF baby. But fears of cloning kept haunting him. "The popular interest in cloning is amazing," he wrote in *A Matter of Life*. "Though it is irrelevant to our work, I have had to face arguments about it since our earliest days."

"Ultimately," one reporter discussing their research wrote, "we could have the know-how to breed these groups of human beings—called 'clones' after the Greek word for a throng—to produce a cohort of super-astronauts or dustmen, soldiers or senators, each with identical physical and mental characteristics most suited to do the job they have to do."

Edwards was even challenged by his brother, Sam. "We call it 'clowning,' not 'cloning,' I told Sam."

Edwards was blasé about cloning because cloning people made no sense at all. Maybe you'd like to clone sheep, say, or cows, because special sheep win ribbons at fairs for their wool, and cows that squirt out big payloads of milk can make a farmer a lot of money. Sure, you'd like to keep those genes around. But people? The reporter's scenario was especially unlikely. "Terrible brave new world visions such as those . . . irritated me," Edwards wrote. "They still do. They are based on the pessimistic assumption that the worst will happen. The whole edifice of their argument is fragile—that nuclear physics led inevitably to the atom bomb, electricity to the electric chair, civil engineering to the gas-chambers. Surely acceptance of the beginning does not necessitate embracing undesirable ends?" Besides, to clone yourself legions of army men, dustmen, and senators, you'd have to make a few other changes to society, like reinstituting slavery and abolishing free elections and if you did those, well then, cloning was the least of your worries. Edwards knew that if you were to clone, you'd end up with a baby. Period. And any way you could think of to have a baby would probably be an easier way than cloning.

In the days after Dolly, just as it had in Edwards's day, much of the world's imagination made the tremendous leap from a sheep in rural Scotland to people. Human clones might walk the earth. Few people seemed to stop and think about why, ipso facto, this would be a bad thing, it just became an article of faith that somebody—Saddam Hussein was often named as a likely suspect—was going to, and when he did, there would be hell to pay because, as New York's Catholic cardinal, John O'Conner, famously declared, you could create a whole cloned army, stick guns in their hands, and send them out to slaughter. The fact that almost none of this made any sense (you'd have to wait, oh, eighteen years for 10,000 babies to grow up and when they did, would they be engineered as mindless drones? If so, how could they be trained? Would the government press gang clones into service?) did not stop people from saying it until, seemingly through repetition, it became an accepted fact. The reality of Dolly, though, and of all clones, was much different, and most scientists knew it.

HARRY GRIFFIN WALKED WITH ME on the little gravel path that leads toward the barns from the main research building of the Roslin Institute. Griffin, a large man whose mass of beard made him look like he belonged in the drizzly, cold Scottish countryside of late March, had made this little trek a couple hundred times, I supposed, over the course of the past five years. Everybody who visited the institute wanted to see Dolly, and, like a worn-out Universal Studios tour guide, Griffin plodded down the gravel, showed off the sheep, and plodded back to work. You might think he would be annoyed but he wasn't. The feeling he had was more akin to amusement. People came to look at Dolly with the same nervous expectation carnival goers get when they enter the freak show tent, testing their conception of themselves against the sight of the hermaphrodite or the lizard boy. Dolly even came equipped with a built-in barker's pitch: Dolly, The World's First Cloned Mammal! What truths would the viewing of her specialness reveal?

Griffin and I passed a stone memorial commemorating a particularly bloody battle between the Scots and the English during the 1300s. They say, Griffin told me, that the nearby burn, or creek, ran red with blood. That was the Scotland of old, when life was short, cold, brutal. It had since been turned into the Scotland of tourism ads, all whiskey and kilts and tossing kabers, and now it was turning itself into a new Scotland with other, newer ads. In every major science journal, Scotland was advertising for scientists and for biotech investment. Dolly was often pictured, groomed and perfect. Dolly was a product of Scotland, and the economic development folks were damn proud of her. She proved that biotech did not have to happen in Oxford or Cambridge, by the shores of San Francisco Bay, or in San Diego or Massachusetts—all the usual centers of biotech.

Griffin and I reached the barn—actually a big metal and concrete shed—where Dolly and her extended family lived in the lap of
luxury, at least by sheep standards. I stepped up to the tubular railing
inside and my own revelation was immediate:

Dolly was an old, fat sheep.

Over my shoulder, I could see Griffin. He smiled just a little and
I was reminded of the reaction of cattle cloners as I stood in a Wisconsin pasture filled with cloned cows. I stood there, looking at the
cows, and waiting for, well, something. But all I saw were big doe-
eyed cows. And now here was Dolly.

She ambled to the railing. Her enormous winter cloak of wool
rolled in seismic waves across her body. She looked very different
from the ads in the magazines. Her wool had accumulated months'
worth of whatever sheep wool accumulates and to me she looked all
the world like a down-at-the-heels dowager whose inheritance had
run out. She struggled to retain her former dignity, stopping directly
in front of me, a foot away from the rail. She stared me in the eye,
virtually demanding that I supply a treat. Visitors often brought
snacks for Dolly, a form of tribute that plumped her into the Imelda
Marcos of the sheep world. But I had nothing to give.

"Baah."

I thought I heard a note of disgust. She showed me her rump
and waddled away, limping from arthritis.

Griffin looked into the pen and sniffed, "Her royal highness."

Griffin's a serious guy but he's got a pretty good sense of humor
and I think he got a kick out of all the fuss Dolly stirred up when she
was introduced to the world, all that "hoo-ha" as Wilmut called it,
the scary science fiction scenarios offered up by media and politicians, the frenzied talk about the copying of people and the cloned
armies and the teams of Michael Jordan DNA knockoffs. But here
she was, just a rather spoiled sheep.

Dolly was boring. Cloning was boring. The Wisconsin cows were
boring. People sure got worked up about cloning, tittered over all the
cool science fiction you could think up in your head, but in the six
years since Dolly's birth, real cloning to make a sheep, or a cow, or a
mouse had become boring. Wilmut knew it, lots of cloners knew it.
Cloning to make a person was probably most boring of all. Why

cloning worked, that was interesting, but cloning for the sake of cloning? Boring.

Despite the worldwide frenzy over human cloning, cloning wannabes, with the exception of the profoundly infertile, were viewed within the biotech parallel universe as sadly misinformed and silly. Even the life extensionists, who had once touted cloning to make bodies for all those frozen heads, had come around. They now thought that nanotech machines would rebuild bodies from scratch. The transhumanists, the neo-eugenic hopefuls, all of them agreed that cloning was a pretty boring concept, merely an endless, pointless genetic samsara that got you no closer to biotech Nirvana than you were before you started. Every mistake would still be there, written indelibly into a clone's genome without hope of perfection, progress, expansion. There was no immortality from cloning, not even a hint. In fact, the mind of a clone would be brand-new, the C drive empty upon birth. You'd have to start all over again, filling it up, rebuilding all that brain power. This defeated the whole purpose of transhumanism, which was to preserve and expand one's data bank, to make bigger and bigger C drives, cram more and more knowledge in a constant Singularity mind blow.

If it weren't for the politics, cloners themselves would not particularly care if people were cloned. Of course, they condemned the idea on the grounds of safety. They knew all too well how frustrating cloning experiments could be and how many times embryos, and then fetuses, and then newborns, died. Nobody knew if that would happen with people, and anyway, as time went on, scientists began to think they would not even be able to make clones of people using the same techniques they used to make Dolly and other animals. Besides, safety provided the perfect alibi for scientists who did not really want to go around telling the world that cloning people would be a meaningless, insignificant development after the world had gone and made up its mind.

Sure, human cloning was going to happen. It may have happened already. Who knew?

But if a baby were to be born through the use of cloning technology, the sun would still rise in the morning. "If the first clone is born it's not going to kill the Earth," James Watson said. Cloning scientists just could not get themselves to believe that babies were scary. Cloning was a straw man and they knew it. Even if human cloning did become safe someday, even if the success rates jumped, and even if it became an accepted part of IVF, cloning to make people would be a lousy business. After all, sex was free and it felt good. You could make babies in the back seat of a 1985 Ford Taurus, at the beach, or between 300-thread-count sheets at a Ritz-Carlton accompanied by a nice merlot and the cool sounds of Dave Brubek. Laboratories would always be a last resort. People who wanted to re-create dead children would be sadly disappointed. Clones weren't copies. Dead was dead. Mass culture was much better at creating mass behavior. There was a Starbucks selling coffee for $2.50 at what seemed like every American intersection. The only people for whom cloning made any sense, the few infertile couples who failed with other IVF procedures, were a very small minority and many of them would probably never want to try it. But so what if they did? Fretting over the ways such parents might hold preconceived expectations for children produced by cloning seemed misplaced as long as dads were beating the tar out of each other at Little League baseball games, babies were baptized into religions, and wealthy New Yorkers were sending four-year-olds to preschools that cost more than many colleges. For scientists, it was hard to see how human cloning made the slightest bit of difference. Dolly did not mean a goddamn thing.

But as James Frazer wrote in *The Golden Bough*, human beings want to make meaning out of the enormity of life and death. Such was the case with cloning. Cloning was just so extraordinary, so fantastic, that there simply had to be something deep embedded within it. That's what mattered to the bio-Luddites.

If people came to believe that Dolly did not mean anything the bio-Luddites would lose the most powerful weapon they had to battle the bioutopian faith. Artificial insemination was bad enough. IVF even worse. But not to need a sperm and an egg? If Dolly did not mean anything, if she were not a portent of disaster and if a cloned human being would be just a baby, then what of the soul? What of embryos? Didn't an embryo mean anything? What about human nature? What about nature, period? There were supposed to be rules, lines of demarcation, so we knew who and what we were. If there were suddenly no rules, or if the rules were meaningless or could be changed, then we would be cut loose from the moral, social, cultural tethers of biology.

Embryonic stem cells were an especially dangerous threat to these tethers. Cloning itself was a dead end, a tool the bio-Luddites could use to stoke biofear. But as biological Play-Dough able to be engineered and morphed into any number of fates, stem cells, the bio-Luddites argued, could allow the human species to create entirely new branches of evolution. The bio-Luddites decided to make a stand, and, with the election of George W. Bush in 2000, they had a new and very powerful ally.

FROM WITTWATERSRAND TO WASHINGTON, scientists, politicians, moralists, and people at truck stops who had never given an embryo a second thought spent much of 2000 and 2001 locked in furious debate over the tiny clouds of cells that Michael West had championed in an attempt to cure aging. Thousands of newspaper column inches, long minutes of TV time, passionate hours of debate on the floors of legislatures were devoted to the issue of ES cells, government funding for research on the cells, and whether cloning technology could be used to create cells for therapy that would be compatible with recipient patients, so-called therapeutic cloning. Advocacy groups rallied around both sides of the issue. For example, CuresNow, a Los Angeles-based lobby, was set up by four movie industry executives who had children with juvenile diabetes. They argued that their children were being denied cures. Right-to-lifers, who believe that we are people from the moment of conception, mobilized too. Their rhetoric reached its pinnacle when an overwrought man named John Borden, who used donated frozen embryos to have two sons, held his sons up before a congressional panel and asked, "Which one of my children would you kill?"

Nobody was considering killing either of Borden's children, but the government was thinking of giving money to scientists so they could dismantle embryos for the stem cells inside, which meant the issue tapped directly into the American obsession with abortion politics. The Catholic Church and evangelical Christians declared ES cell research to be murder.

Under those circumstances, and with the political heat already on, biology was woefully unequipped for the battle. Few scientists are good at, or much interested in, politics, Haseltine being one important exception. For the most part, scientists avoid politicians. Describing the early opposition to IVF, Florence Haseltine, one of its early practitioners, said, "I think it was only politicians yelling and screaming about something they did not know anything about. I did not think about it very much. I thought about ways to get things done."

Scientists rather naively assume that if one presents enough data in support of a theory, the theory will be adopted as sound, at least until some other, better, theory takes its place. This is not how politics works and it is not how politicians behave. In politics, perception, not facts, are key and complexity is something to be avoided. So biologists, with their godlessness, their nuanced intellectualism riddled with caveats, their citizen-of-the-world collaborations, are viewed with suspicion among politicians, a suspicion enhanced by their occasional left-wing outbursts, like Linus Pauling's antinuclear activism and the Vietnam War protests. So biologists try to keep their heads down. For most of the years since World War II and the days of Vannevar Bush, this worked. The politicians did their thing and did not delve too deeply into the work of the scientists. They gave the scientists money, the scientists cured polio and smallpox and fought the war on cancer and assumed, rightly, that the politicians did not particularly care about the intricacies of restriction enzymes, say, or cellular receptors. Everybody was happy.

The system did sometimes break down and when it did, it reminded both sides of why they so rarely spoke to each other. Now, almost everything that had occurred since the creation of Dolly and the isolation of human embryonic stem cells had only reinforced the idea among scientists that politicians, pundits, and most journalists were not to be trusted. Human embryonic stem cell research elicited cries about "baby farms" (and, of course, Brave New World) from politicians and the chattering classes. Some freaked out, warning that science was moving far, far too quickly, that we were about to begin designing genius babies who could deliver nine innings of perfect sinker ball pitching between gigs as underwear models equipped with washboard abs and blue eyes. It was always blue eyes, for some reason, and blonde hair, too, evoking a near-term world populated by Icelanders.

Conservative columnist and leading bio-Luddite Charles Krauthammer hyperventilated in the pages of *Time* magazine. "What really ought to give us pause about research that harnesses the fantastic powers of primitive cells to develop into entire organs and even organisms is what monsters we will soon be capable of creating." Then, referring to work done at West's Advanced Cell Technology (he was not aware that such experiments had been done for nearly a decade), Krauthammer fretted, "In 1998, Massachusetts scientists injected a human nucleus into a cow egg. The resulting embryo, destroyed early, appeared to be producing human protein, but we have no idea what grotesque hybrid entity would come out of such a marriage. Last October, the first primate containing genes from another species—a monkey with a jellyfish gene—was born. Monkeys today. Tomorrow humans." Don't believe the scientists, Krauthammer said. They were sure to take these experiments all the way to people, no matter how many times they denied it. "They will seduce us into forging bravely, recklessly ahead. But just around the corner lies the logical by-product of such research: the hybrid human-animal species, the partly developed human bodies for use as spare parts, and other grotesqueries as yet unimagined."

Krauthammer may not have realized it, but the monkey with the "jellyfish" gene was not quite as freaky as it sounded. Scientists in Oregon took primate stem cells and engineered in a commonly used gene for "green fluorescent protein," a lab tool that allowed scientists to observe cells and to see which of them took up the gene. Scientists had been doing this in lab mice for years. Krauthammer's fear of "human-animal species" was also a tad late. Thousands of "human-animal species" already existed. Every mouse that carried a human gene was a man-animal chimera of sorts, and, for that matter, so was every one of the billions of bacteria pumping human proteins out of vats in biotech company production facilities.

But the bio-Luddites won the day and congressmen started writing anticloning laws. Cliff Sterns of Florida introduced one such bill. "I think it is playing God and interfering with the natural order of creation," Sterns said. For one thing, a clone would be a freak. Who would the father be? The mother? And besides, cloning would start a new eugenic movement that could lead to categories of people because I think in the end, a person without a father or mother, a parent, or a family will have a different identity which might be categorized into scientific categories which has an implication in terms of the ethical ramifications. Does a person that comes from cloning have a different consciousness? A different relevancy? If you meet them at a social or professional level, if you knew they were cloned, how would you deal with them?

And we would know a clone when we saw one, too, because, Sterns insisted,

When you do a clone, there are these tentacles, part of the ovum. They remove that. There's an actual term for it. When you clone, you do not have an exact clone of the ova material. The tentacles are all removed . . . The clone would not have these and yet you and I have these when we are born. If we clone ourselves, we would not have them. We would have a category of somebody, people who did not have these tentacles and these might be superior or inferior people.

This was the sort of explanation that made scientists bury their faces in their hands, speechless. But such misconceptions were popular. On April 14, 2002, pundit George Will appeared on ABC's "This Week" with George Stephanopoulos and argued that all forms of cloning, therapeutic or not, should be banned because "these are entities with a complete human genome." In fact, just about every cell, red blood cells being one exception, has a complete human genome. By Will's logic, you could not tamper with any cell in the body, even cancer cells.

Not all conservatives relied on their gut. *New York Times* columnist William Safire, for example, engaged scientists in discussions. During a meeting of the World Economic Forum in Davos, Switzerland, Safire, Wilmut, neural stem cell researcher Fred Gage, Stanford stem cell scientist Irving Weismann, and renowned cell biologist Bruce Alberts got into a funny conversation about how many human neurons you could stick into a rat brain before the rat wound up with a human brain. A rat with a human brain was crazy to think about (a Warner Brothers cartoon show called "Pinky and The Brain," about two lab mice equipped with human brains and bent on world domination, used this device), but the real answer, the scientists explained, was probably that the rat brain would use the human neurons like rat neurons and you'd just wind up with a rat carting around a lot of extra brain cells. Safire later wrote a column urging Bush to be bold and fund the research.

But such nuance escaped many others who, like Krauthammer, opted to smear scientists. Three Republican leaders of the U.S. House of Representatives labeled biologists executives in "an industry of death." Biologists were referred to as "mad scientists" on the floor of the U.S. Congress.

Roger Pedersen finally had enough. The Labour government of British prime minister Tony Blair had not only endorsed ES cell science, including cloning technology, but was hoping it would become a big economic force. So, during the summer of 2001, Pedersen picked up and moved from San Francisco to Cambridge University, instantly making himself the poster boy for what ES cell advocates warned would be a U.S. brain drain, the start of an expatriate colony of biologists.

Douglas Melton, the chairman of the department of molecular and cell biology at Harvard, thought he could do something about all this. He thought that if he explained the science, advertised the potential, the politicians would be moved and see that biologists were not, after all, madmen. A year after his experience, he admitted that he had been naive. "I have learned the hard way I am not very good at politics," he said.

Melton is a thin man, with dark, curly hair, and he wears a pair of big glasses that make him look very much the way everybody thinks a biologist is supposed to look. Normally, he is soft spoken. But he's fit and wiry and clearly has not spent every moment of his adult life over a microscope. You can tell this because when he talks about his science and his run-in with politicians, the muscles in his neck and shoulders tense like bow strings.

In the early spring of 2001, when it became clear that Bush was trying to walk on the balance beam between the people who were desperate to use ES cells for all the problems scientists had said they could be used for—Parkinson's and Alzheimer's patients, the paralyzed, the organ-needy, diabetics—and the people for whom an embryo was a living person—the Catholic Church, evangelical

Christians, certain members of Congress—Melton became a leader among biologists in attempting to persuade the president to tip on the side of science.

Melton was originally drawn into the stem cell debate because he had contact with the office of Massachusetts Senator Edward Kennedy. Melton sometimes advised Kennedy's staff on biology matters and when it became clear that the politics of ES cells was coming to a head, Kennedy's office arranged for Melton to meet with Senators Arlen Specter of Pennsylvania and Diane Feinstein of California. That led to meetings with Health and Human Services Secretary Tommy Thompson.

Thompson was regarded by biologists as their man in the Bush administration. He was the former governor of Wisconsin, home to the University of Wisconsin, which had a rich history of molecular biology. It was also, of course, the place where human ES cells had been isolated and successfully cultured for the first time. Its ag animal lab rivaled the one John Hammond had set up in Cambridge, England, and had been home to some of the most revolutionary experiments in farm animals. A Wisconsin company, Infigen, created the first cloned bull, "Gene," and as governor, Thompson actively promoted the biotech industry in his state.

Sure enough Melton found Thompson receptive. To Melton's ears, the secretary seemed enthusiastic about having the NIH, which he oversaw as part of his department, pay for human ES cell research. Melton explained the bridge biologists wanted to build between the reductionist century and the century of the cell. We were, he said, on the precipice of a great new age. Thompson agreed that this was indeed the frontier of biology. "I was left with the impression that if it was his decision to make, the NIH would have been charged, as Tony Blair had charged Britain, to make sure the United States was the best place to do stem cell research. I honestly believe that was Tommy Thompson's view of the world." But Thompson, and Thompson's office, were not the obstacles. The White House was.

Because Melton has a son who is afflicted with juvenile diabetes, he is active in the Juvenile Diabetes Research Foundation (JDRF), one of the most powerful and connected of the patient advocacy groups that lobbies in Washington and supports research. An old friend from Bush's Yale days, who also has a diabetic son, managed to work the JDRF into the White House. The organization asked Melton to be the face of science in the meetings.

First he met for about forty-five minutes with Vice-President Dick Cheney, who seemed genuinely interested in the issue. Cheney asked penetrating questions about stem cells, but he also told Melton that since the president was going to make the call, Melton should meet with Karl Rove, the president's political director. Melton admitted that he had never heard of Rove, famous for being the political whiz who had engineered Bush's election and for being Bush's link to conservative Christians.

On May 10, Melton went to the White House to meet with Cheney, Deputy Chief of Staff Josh Bolton, and Rove. Bolton and Rove were just too busy to hear or understand much of what Melton tried to tell them and anyway, Rove appeared to have already made up his mind. "He said, 'Even if you had stem cells working, you'd always have immune rejection problems so why bother,' or something like that," Melton said. "That was sort of silly. Nobody was saying this solves all problems. It's complicated.

"He gave me the impression he was trying to find reasons not to be supportive of this work . . . the discussion did not go in a way that [suggested he was] trying to determine what was the right thing to do for health and human services, it was the question of 'What is the political calculation about our decision?'"

A month later, Melton and the JDRF met with Bush. Rove was there, again, and so was Chief of Staff Andrew Card and Bush's longtime operative, Karen Hughes. Bush, though, seemed unengaged. He had just come from a quick meeting with Mel Gibson, who was promoting his new movie, *The Patriot*, and while Melton's meeting was not a photo op, Melton was left to wonder what the point was. "It was like, 'Here's your chance to tell me something, so I'm listening.'" Melton simply stressed repeatedly that human embryonic stem cell research had nothing to do with abortion. Embryos in dishes were not people. They could never be people.

On August 9, 2001, the president made his now-famous televised address to the nation on stem cell research. Speaking from his ranch in Crawford, Texas, Bush tried to split the difference between the two sides of the debate, but it was clear he had been much more influenced by Kass than scientists like Melton. He described his journey toward the decision, saying he gave the issue "a great deal of thought, prayer, and considerable reflection." But, he told the nation, "We have arrived at that Brave New World that seemed so distant in 1932 when Aldous Huxley wrote about human beings created in test tubes in what he called a hatchery." ES cells were "at the leading edge of a series of moral hazards," like "growing human beings for spare body parts." ES cell research, he said, offered great promise but at great peril. In light of all that, Bush said he would allow the feds to pay for research on "more than sixty diverse stem cell lines" that already existed, but no more. He also directed the formation of a bioethics advisory council to be headed by Leon Kass. On that day, bio-Luddism became enshrined in the federal government.

Many scientists tried to sound relieved that Bush agreed to any funding, but it soon became clear that there was no such thing as sixty diverse stem cell lines available for scientific research. Nobody was sure just how many there were, but it seemed likely that you could, if you wanted to, count them on one hand, two at the most. And anyway, the most certain candidates, the ones scientists knew the most about, were the ones made by James Thomson using Geron's money. Those cells were tied up with patents and Geron licenses, and came with complex strings attached. In effect, Bush had effectively stopped any large-scale research on human ES cells with federal dollars.

Since the federal government is the biggest funder of basic biology, spending about $14.6 billion for research grants to scientists from the NIH budget alone, American graduate students, the real labor force of biological research, began avoiding ES cell work. The students simply had no prospects for career advancement in the United States. With his Brave New World imagery, Bush managed to lay a film of unsavory suspicion over the entire field. It didn't matter that no scientist had ever proposed "embryo farms" and growing human beings for spare parts. The first organization to receive funding under the Bush rules, the American Red Cross, was so wary it wound up turning the funding down, even though it had applied for it. The Red Cross denied it was trying to avoid controversy, but biologists did not believe them.

"They are worried about [blood and financial] donations!" Melton was furious now. "You could make blood! Not suck it out of people!"

Bush also wanted to criminalize therapeutic cloning, the use of nuclear transfer to make stem cells as West and then Geron had envisioned it. He said he supported a bill proposed by Senator Sam Brownback that would not only send scientists to jail if they tried it, but would also outlaw treatments developed in other countries that resulted from the use of cloning technology. The prospect so alarmed biologists that Paul Berg, UCSF chancellor and Nobelist J. Michael Bishop, and Intel founder Andrew Grove took out a large ad in the *New York Times* begging Congress not to pass such a law.

Science also tried to appease critics by repeatedly telling anybody who would listen that they, too, believed embryos weren't just a bunch of cells, that embryos had a special status of some kind. Some used the word *reverence*. Most biologists, especially developmental biologists, didn't really believe this, at least not in the way nonscientists and especially Catholics and evangelicals meant it. Biologists did find human embryos awe-inspiring, but they found chick embryos and mouse embryos awe-inspiring, too. Experimenting on one was no different than experimenting on another except that human embryos were more difficult to obtain and came freighted with a lot more baggage. Even scientists regarded as conservative saw nothing wrong with experiments like mixing human cells and cow eggs in an effort to retrieve human stem cells. "What I guess it really means," one explained, "is that you are not respecting the embryo, you are trying to either respect or assuage those people who do not think you should be using embryos for research at all."

No amount of verbal tap dancing assuaged religious conservatives. In May 2002, journalists received an e-mail from Panos Zavos, a formerly obscure Kentucky fertility doctor, who had said he was going to start cloning people. He had been asked, he said, to testify before the subcommittee on criminal justice of the House Committee on Government Reform which was looking into reasons why therapeutic cloning would make cloning for babies that much more likely. Zavos was not interested in making stem cells. He wanted to make babies, or so he said, and he had already testified before Congress on that topic. He was certainly excited about going again.

Zavos had signed with the same big-time speakers bureau, APB, that also handled Larry King, Ralph Nader, and Desmond Tutu, and every time he showed up on TV, his Q ratings rose.

Roland Foster, a committee operative, was organizing the session. Foster had been an aide to representative Tom Coburn who, before his retirement, had sponsored a string of socially conservative bills calling for sexual abstinence, the banning of the abortion drug RU-486, and the registration of people with HIV. Now Foster was an aide to Indiana Republican Mark Souder. He knew that most people following cloning thought Zavos had no chance of cloning anybody and was, mostly, a media hound. But he laughed at the prospect of Zavos's testimony. He wanted to use Zavos for political theater, as scary guy du jour. This was Washington, and the hearing was about politics not science.

"Before going" to Washington, Melton said, "I thought part of the reason for me bothering to go on my own money, to fly there to talk, is that they may have genuinely wanted to listen. I now would be slightly more cynical."

Melton's story illustrated how the cloning and stem cell debates led to even more disaffection among scientists. Biologists deeply resented the accusations that they were somehow less moral than congressmen or columnists or the bio-Luddites. The scientists had wives and husbands, golden retrievers, mortgages, and cars they'd like to trade in for a new Audi. They almost never laughed demonically. And they had a long history of policing themselves, most famously at Asilomar. Some scientists, like gene-therapy pioneer Theodore Friedmann, spent about as much time pondering the ethics of their work as they did actually doing it.

"What does it take to be a bioethicist?" Melton asked rhetorically. "What would it take? My undergraduate degree is in philosophy and when you read a book by Wittgenstein, you know you have read a work by somebody who has thought deeply about something. This business about announcing you are a bioethicist and you will comment on what is moral and what is not, which real philosophers would never do . . ." Melton was worked up. He was on a roll. The tendons in his shoulders and neck popped out of his T-shirt. "Why can't scientists comment on bioethicists! Kass and his buddies pretend and present themselves as real moral authorities! The arrogance of this is shocking!" The president's bioethics commission, popularly known as the Kass Commission, made "idiotic, laughably stupid statements."

The commission, which included Kass, Fukuyama, and Krauthammer, among others, was perceived within science as so heavily weighted with bio-Luddites that its conclusions were a given. During the very first meeting, in January 2002, Krauthammer went beyond the most optimistic Extropian and insisted that scientists were just about ready to create "a class of superhumans." To open the meeting, Kass led a discussion of Nathaniel Hawthorne's short story, "The Birthmark," in which a scientist tries to erase the mark from his bride, an already lovely woman. In the process he destroys her beauty.

Science rolled its eyes. "That, and his support of human instinctive distaste as a fundamental moral measure of new developments," editorialized *Nature*, "suggests a determination to confront the research agenda not only with ethical discussion but also with irrational fears and pessimistic foreboding."

As if to hammer the final nail in the coffin of the Bush administration's credibility within the largely agnostic biology community, Bush greeted the members of the panel at the White House by telling them the commission would not only address the coming biotech quandaries but that it would promote the idea "that there is a Creator."

Nine months later, the commission issued a majority report recommending that the government impose a four-year ban on any research that involved cloning technology. Four commission members, all scientists, issued a resounding dissent in *Science*: "That proposal is short sighted. It will force U.S. scientists who have private funding to stop their research, and it will accelerate the brain drain to more enlightened countries."

In fact, there was something of an ES cell and cloning land rush going on around the world. China had set up biologists, many of them trained in the United States, in state-of-the-art labs. Scientists there began inserting human cells into rabbit eggs to make cloned ES cells. Japan established the Center for Developmental Biology in Kobe in an effort to make Japan a world leader in the field. The new Japanese center's first big recruit was Teruhiko Wakayama, a cloning whiz who worked for West before moving back to Japan.

In India, Reliance Life Science, a division of a huge industrial conglomerate there, not only possessed its own ES cell lines but was aggressively recruiting scientists to turn them into commercial ther- apies like tissue-engineered organs and cells for diseases like Parkin- son's. In Singapore, home to ES Cell International, the company that spun out of an Australia–Singapore team's own successful derivation of ES cells, the government committed hundreds of mil- lions of dollars to build a city within the city called Biopolis. It approved the use of cloning technology to obtain stem cells, then began advertising for scientists. Johns Hopkins's Singapore Biomed- ical Center, an affiliate of the Baltimore university where John Gearhart pioneered stem cells, was one of the first research groups to announce it would move in. Sweden, Israel, Australia, and Korea all also announced major stem cell initiatives.

Still, the brain drain argument was probably oversold. Most American biologists simply took their disgust as another object les- son on why it's best to have as little as possible to do with politicians and moralists, why, in the words of Haldane, there could be "no truce between science and religion." It was time to retreat back to the labs and figure out how to keep working.

Sometimes they could cooperate with researchers in other coun- tries. Sometimes they could find private money given by people who said, "If the government won't fund it, we will." Andy Grove gave $5 million to start a stem cell center at UCSF, the place Roger Pedersen had left to go to Cambridge. An unnamed private donor gave Stan- ford University enough money to set up its own ES cell center. Advo- cacy groups like the JDRF, the Parkinson's Foundation, and the Alzheimer's Foundation gave money. As exorcised as Melton was about his experience in Washington, he was able to derive his own stem cell lines using money from the Howard Hughes Medical Insti- tute and donated embryos obtained from a Boston IVF clinic. Money from the NIH, the biggest biomedical funding agency in the world, would certainly have helped. But ES cell research could limp along without it.

SCIENCE MAY HAVE ROLLED its eyes at the bio-Luddites, dismis- sively denied their sci-fi scenarios, but the bio-Luddites were right about the rapture. Scientists, to say nothing of transhumanists, could not deny that many of the bio-Luddite fears might actually come to pass. As far as biotech was concerned, there were at least 1,500 good reasons to rewire human biology. That's how many diseases, at least, science reckoned were caused by some balky gene or combinations of genes. In fact, when you stopped to think about it, we were pretty messed up. Yes, we'd made it through four million years of evolution, but we'd picked up a lot of garbage along the way. Every person's genome had some thing or other wrong with it. White Europeans suffered from cystic fibrosis, their lungs filling with putty-like mucus, their bodies starved for breath. Africans had sickle cell disease, a mutated gene that turned their red blood cells into tiny boomerangs barely able to carry oxygen. Italians and Greeks and Cypriots had thalassemia; Jews had Tay-Sachs. There were hairlips, Down's syndrome babies, fragile X. People were born with genes for dozens of "syndromes" like Marfans, Kleinfelter, Rett's, Cri du Chat, Wiscott Aldridge, Kartageners, Pelizaeus-Merzbacher, Leigh's. There was myotonic dystrophy and cherubism and neurofibromatosis. And good god, the cancer genes! Evolution, James Watson said, "could be damn cruel."

All these twisted genes were the lint of the ages glomming onto us with the static electricity of happenstance. Despite the publicity, there was no such thing as "a human genome," or one "Book of Life," as human genome sequencers liked to call it. Everybody had their own genome. There were billions of books and they all carried misprints. Genomes were not nearly as fixed as we wanted to think, as immutable as pop culture had come to believe. Evolution goes on. Genomes change. Some of the DNA we cart around right now came from viruses, from bacteria, from the mating of our parents. "We are the last five minutes of evolution," Robert Edwards liked to say, "and we have gone wrong! Badly wrong!"

Many people—people who were not transhumanists, had never heard of Hans Moravec or of The Singularity—already clamored to rid their families of genes gone bad. In 1963, Edwards was part of a team that analyzed a rabbit embryo and was able, by looking at its chromosomes, to determine the rabbit's sex. Twenty-six years later, in 1989, two English scientists were able to do the same for a human embryo, a development a British politician called "an affront to human dignity." But by late 2002, over 4,000 embryos had been tested using a technique called preimplantation genetic diagnosis (PGD) to detect dreaded mutations, and the numbers were growing exponentially. Often, the parents requesting PGD were not infertile. Yet they went through all the rigors of IVF so their embryos would start growing in a dish, not a uterus. That way, a cell from each embryo could be removed. The DNA in the cells could be tested for known mutations carried by the parents, and not just deadly childhood diseases like Tay-Sachs, either.

Doctors saw a lot of patients who wanted to know everything possible about genetic testing, about PGD, about embryo selection. They heard story after story from devastated, heartsick mothers and fathers who not only inquired about, but begged doctors to use every available technological option to erase their particular genetic misspelling from their family's DNA dowry. Mothers who had a history of breast cancer in their families were having embryos tested for mutations related to breast cancer, a disease that would not strike until sometime in adulthood. Parents whose relatives died of other cancers were testing embryos for an anticancer gene called P53.

"Preimplantation genetic diagnosis has recently been offered for couples with inherited predisposition for late onset disorder," began the abstract of a 2002 journal paper by a leading PGD expert. His group used the procedure to test for risky cancer mutations. In this paper, the biologist's data covered "twenty PGD cycles performed for ten couples, resulting in preselection and transfer of forty mutation-free embryos which yielded five unaffected clinical pregnancies and four healthy children . . . Despite the controversy of PGD use for late onset disorders, the data demonstrate the usefulness of this approach as the only acceptable option for at risk couples to avoid the birth of children with inherited predisposition to cancer and have a healthy child."

PGD was increasingly being used for nondisease traits, too. Parents who had a child who needed treatment for leukemia, Fanconi's anemia, and other diseases requiring blood-making cell transplants, were testing embryos to make sure their next baby would make a good donor. A few parents were testing embryos so they could choose the sex.

Choosing a trait like sex made a lot of people nervous, but not Edwards and not Joe Leigh Simpson. Simpson, a Baylor University geneticist and a past president of the American Society for Reproductive Medicine, told his fellow IVF doctors at their annual convention in 2001 that if parents wanted a daughter, or if they wanted a son, why should a doctor stand in their way? "Why are we so frightened about it? After one or more children of a given sex, what is wrong with letting couples select the desirable sex in the next pregnancy?"

Edwards, a father of IVF, the winner of his own Lasker Award, agreed. He was not concerned that parents would choose boys over girls, the usual fear expressed by critics of the idea. If an imbalance developed, he said, "we'll stop making the popular sex for six months until the other sex catches up again." While Edwards's view was a minority opinion, it was not radical within the world of IVF and PGD. Long and sometimes impassioned e-mail debates among PGD experts were held over the propriety of choosing the sex of an embryo. Some doctors argued that they knew of too many cases involving fetal testing, ostensibly for a condition like Down's syndrome, whose real purpose was to discover the sex of the fetus. The undesired answer was often followed by an abortion. Why not choose the sex the parents want from the get-go?

PGD rose to satisfy these new markets and, in 2002, formed its own international society. The official journal would be Edwards's *Reproductive Biomedicine Online*.

To Simpson, the world science was creating gave parents the power to choose the genetics of a future child. The consumer was king. "We are here in Orlando," the home of Disneyland, he said at the 2001 convention, "and I would suggest we too have our Magic Kingdom. Hospitals, clinics, laboratories, that is our setting, our environment. And just like the Magic Kingdom, I would submit we are cast members. We have responsibilities to pursue scientific inquiry as well as optimal patient care. We have the responsibility and obligation to conduct research to the best of our ability and to develop new technology and always reject the status quo." The Imagineers in Simpson's Magic Kingdom could do far more than choose sex or select embryos unaffected by fragile X or sickle cell. "In doing so, we should listen to the guests, to our own patients. We should not be paternalistic and assume we know what they need . . . If we do that, we will not have any trouble with the new technologies, the new genetics. We have every reason to embrace it and to encompass it, and not to fear it."

Simpson, and those who thought as he did, wanted to break free of what he considered "pre-Reformation" squeamishness about fixing nature's mistakes. What was the difference, say, between giving a kid insulin for the rest of his life, and inserting an insulin-making gene into an embryo that lacked a good one? Same went for any genetic disease, especially the single-gene disorders like cystic fibrosis, caused by one bad apple. Not only would the kid be cured, but he would not pass the bad gene on to his kids nor they to theirs. They would not even be carriers of one copy of the gene that, when paired with a copy from another carrier, would make a sick child. No need to test potential mates. No need to worry that a time bomb was ticking away in the gonads of your girlfriend. You could defuse it. Forever. PGD, he said, had delivered the "Holy Grail."

This kind of treatment with genes is called germline therapy because its effects are felt in the cells that make sperm and eggs and so the changes are inherited. The very mention of germline therapy brought many, and not just the bio-Luddites, to their feet. Very few mainstream scientists would argue for it, at least not publicly. One, Gregory Stock, the director of a University of California at Los Angeles program on technology and society and a transhumanist hero, began agitating for its consideration in the mid-1990s, and wrote a book, *Redesigning Humans,* advocating it. Stock's promotion of the idea helped break the ice, a little. And now Simpson said it was time to start experimenting with embryos to deliver permanent cures.

After all, as he and Edwards argued in an editorial in Edwards's journal, success breeds acceptance. "Most ethical disagreements have been vaporized by the pragmatism of 1,000,000 IVF babies and 1,000 PGD babies." This was exactly the fear of the bio-Luddites, that facts on the ground would overtake the cautionary warnings. But Edwards and Simpson saw the new genetic technologies as part and parcel of the same old arguments Kass had been raising for thirty years. "The vehemently intense objections made against the original [IVF] researchers and clinical studies dissipated," they wrote, "even while some of the same ethicists who made such misjudgments in the 1970s to 1990s are rising, phoenix-like to attack newer reproductive

advances, often in the same hypercritical mode." Simpson and
Edwards recognized that they were in the minority on subjects like
germline modification, but they didn't think they always would be.
"The current minority toward certain reproductive technologies could
become the majority once children of the newer technologies are
found to be as normal and happy as those conceived *in vivo*."

That was probably true, because while the idea of germline alter-
ations was the nightmare scenario to all wings of the bio-Luddite
movement, you didn't get much argument about it from people who
carried genetic diseases. Nancye Buelow was forty-nine, and she figured she had, if she was lucky, another decade, maybe fifteen years,
to live. Nancy suffered from Alpha 1, a gene-based enzyme deficiency
that results in damage to the lungs and liver and then, eventually,
death. One version of Alpha 1 strikes in childhood and most often
affects the liver. Another version arrives with adulthood and primar-
ily affects the lungs. Nancye was a forty-year-old North Carolina
mom when she was diagnosed and practically from the day she
heard the news, she became an activist. She became a president of
the Alpha 1 Association; she got involved in the Genetic Alliance, an
umbrella group for people with all kinds of genetic diseases; she
served on an NIH citizens advisory panel. Some people in those
organizations opposed embryo testing on religious grounds. Others
were opposed to any kind of genetic manipulation like germline therapy. Not Buelow. If she had known what awaited her when she was
having her children, if such testing had been possible then, she
would have leapt at it. "I sure would," she said. "I am pro research. I
want them to look out for the coming generations, my children and
my grandchildren."

But the bio-Luddites fretted over the "slippery slope," which,
along with "Brave New World" was the other frequently invoked
cliché. They insisted that such testing, any technology to create an
inherited change, was just a razor-thin step away from making super,
enhanced people. This, of course, was what transhumanism was all
about. Bioutopians liked the idea of PGD because it confirmed their
belief that technology was rising to meet their hopes. But PGD was
only a half measure, a quality control step for mistakes. The
"healthy" embryos were not really, truly healthy. After all, there had

been a time when a baby was thought to be born healthy if it was not obviously sick with jaundice, or a tumor, or some deformity. But molecular biology had now shown that a baby who looks healthy, who acts normally, who grows into a smart third grader who's a whiz at phonics, may not be fine at all. He could have a time bomb tick, tick, ticking away.

To the bioutopians, "normal" was not healthy. Normal meant you'd get cancer, or heart disease, or Alzheimer's someday. Even fixing the bad genes, the way Simpson and Edwards wanted to do with germline therapy, didn't really solve the problem. As terrific as it sounded, and as welcome as it was, eliminating cystic fibrosis from a family did not increase their brain power, extend their life span, make them stronger.

Just as the bio-Luddites feared, transhumanists wanted enhancement. "Who can blame us for wanting more and better?" Robert Ettinger wrote in the 1989 preface to *Man Into Superman*. "We think life is great already, but that doesn't mean we settle for what we have." "There are super mice and patented animals and plants resulting from genetic engineering. Scientific heavyweights think that even within your natural life time we can have mental prostheses—direct computer links to your brain. The future is looking good." The bio-Luddites argued that any enhancement somehow defied nature and therefore compromised human dignity.

But beyond this vague notion of the unnatural, the bio-Luddites were at something of a loss to explain why, exactly, the idea was so insidious. Not that they didn't try. They wrote books and editorials and even invented science and psychology on the fly to convince everyone of the danger. But it never seemed to amount to much more than handwringing over a creepy feeling. It was scary, they said, to contemplate letting go of the moorings of our biology to float into an uncertain future. Fukuyama talked of the need to preserve the "factor X" of human nature. He wasn't sure what that was, exactly, but he was sure it existed and was threatened by biotechnology. Kass had his famous "wisdom of repugnance." But nailing down any rational explanation could be, well, a little messy. During a meeting of Bush's bioethics advisory council, member Paul McHugh "looked for examples of what are the problems seen here, and the problems that come from this issue of enhancement."

But it was clear that despite his search McHugh still found the whole thing pretty confusing, though he spent a lot of verbal capital trying to explain it.

> I'm concerned we're talking about a growing break between the higher culture and the common culture in our world. That's affecting medicine as well as everything else. . . . Frank Sinatra is different from Elvis Presley. Why is that? We'll start with Sinatra first. For Sinatra, Sinatra showed us—Sinatra's part of the common world, part of the common culture. He sings common songs, but he shows us [how] wonderful music can be, and that's why we love him. It's not just that bobby soxers screamed about him. I mean, I'm not a bobby soxer and I still listen to Sinatra. Why do I do that? I do it because he makes us aware of just what the music can do.
>
> Elvis on the other hand, the music was to make Elvis. That's why people say Elvis isn't dead, because he's an icon, okay? Elvis lives because—Sinatra is dead because he's a human and made music what it could be.
>
> Elvis is alive because he's an icon and tried to make music make him. You don't sing many Elvis songs. You know, it's the same thing. You could go on in the music area. Louis Armstrong versus the Rolling Stones, Ella Fitzgerald versus the Supremes . . .
>
> The same thing in baseball . . . [Pete] Rose is another, for him, he could gamble, he could do anything, he could debase the game for his purposes, for all that he was—the game was for him and to make him what he is. I can go on a little further.
>
> Medicine is intended to help us overcome an illness and to be what we are intended to be. To use it in another way is to take what we've discovered and misuse it. . . . We need to be sure that we're telling people what they're losing because the higher culture ultimately is to bring us on—we might have to give something up to be in the higher culture, but it makes up for it in all kinds of ways.

Huh? How about the great Bing Crosby versus Sinatra controversy, or the singability of The Supremes' "Baby Love" versus Ella's "Things Ain't What They Used to Be"?

That was really the problem for the bio-Luddites. Things weren't what they used to be. Elvis outsold Sinatra (who had outsold Crosby, who had outsold Eddie Cantor . . .). Until the 1970s, people were not able to conceive babies in dishes. Now they could. People could not even think of changing their genes. Now they could. Life span was immutable. Now a few had begun arguing that maybe it wasn't. The orderly arrangement of the human condition was at risk of being overturned by a bunch of unregulated, white-coated, elitist scientists and their fringe Extropian cheerleaders.

And the bio-Luddites gave a lot of credence to such transhumanists who, until the bio-Luddites got hold of them, amounted to just over a couple thousand people who spent a great deal of time preaching to the converted over the Internet, reading science fiction and science journals, and organizing get-togethers disguised as conferences. In their eagerness to create a scary enemy for the great mass of people who were neither bioutopian nor bio-Luddite, the biological conservatives created a looming transhumanist menace. In fact, opined an article in the conservative *National Review*, the transhumanists were "the next great threat to human dignity."

But with their frequent panicked warnings about the coming rapture, the bio-Luddites wound up giving credibility to the most optimistic bioutopian predictions. Humans were about to be augmented, immortality was around the corner, babies were soon to be designed like Donna Karan dresses. Krauthammer seemed convinced that B-movie monsters were on the way. "Genetic science is advancing at mach speed," wrote the *National Review*.

This was, naturally, intensely unsettling. So to combat the idea, the bio-Luddites placed themselves in the position of the abbey monk Jorge, the library master in Umberto Eco's novel *Name of the Rose*. Jorge kept a book of Aristotle's writings hidden, killing any who dared read it. He thought it dangerous to read because "the book of Genesis says what has to be known about the composition of the cosmos, but it sufficed to rediscover the *Physics* of the Philosopher to have the universe reconceived in terms of dull and slimy matter . . . But if this book were to become—had become an object for open interpretation, we would have crossed the last boundary."

Jorge knew the consequences of crossing the last boundary. "To the villein who laughs, at that moment, dying does not matter: but then when the license is past, the liturgy again imposes on him, according to the divine plan, the fear of death. And from this book there could be born the new destructive aim to destroy death through redemption from fear." This, argued Jorge, would lead to the "acceptance of the base" (like, say, Elvis). "This book could prompt the idea that man can wish to have on earth (as your Bacon suggested with regard to natural magic) the abundance of the land of Cockaigne."

As R. U. Sirius said, the coming rapture was a work of the imagination and imagination can make what were once fringe concepts seem not so fringe anymore. The bio-Luddites feared that Jorge was right: "When what has been marginal would leap to the center, every trace of the center would be lost. The people of God would be transformed into an assembly of monsters belched forth from the abysses."

Thanks to Kass's strategy of turning cloning into a straw man, and setting up transhumanists as a frightening, cult-like force, many people were convinced that monsters were coming, that the biotechnological agenda placed people at risk of dehumanizing themselves. After all, that was exactly what the transhumanists said they wanted to do, to stop being plain humans. The way to prevent that catastrophe, argued the bio-Luddites, was to keep ES cells, cloning, germline therapy, any form of enhancement so far on the fringe it never had the chance to leap to the center. The bio-Luddites succeeded, at least temporarily. But they seemed not have reckoned with the ways science, and the human nature they exalted, are able to defy those who would contain them.

8 WATER INTO WINE

SUNDAY, DECEMBER 2, 2001

Leading the charge toward everlasting life and the elimination of human misery can take it out of you. But as hectic as the past few hours had been, Bill Haseltine was energized. He began hoping for days like this during the 1950s when he was a boy back in China Lake. And now here he was at fifty-seven, at the tail end of a very long day that was all about delivering human beings from suffering. He was about to host a dinner party during which the topic of conversation would be how to save the continent of Africa from the plague of AIDs. He had begun the day in downtown Washington, D.C., at a conference he had organized. Haseltine's meeting, the Second Annual Conference on Regenerative Medicine, was an installment in what he hoped would become a movement to reject the bio-Luddite fear-mongering and replace it with a grand vision of human possibility.

Most of the people about to arrive at the Georgetown home Haseltine shares with his wife, Gale Hayman, were not fully aware of this side of Bill Haseltine. To most of his guests, Haseltine was simply a very smart biologist who was now CEO of Human Genome Sciences, one of the biggest biotech winners to emerge from the gene revolution. They were not science geeks, most of them, they were policy geeks, diplomats, think tankers, aides to officials who did not understand the intimidating language of biology and biotechnology. Haseltine did not routinely drop words like *immortality* among the uninitiated so they arrived focused not on the meeting Haseltine was running downtown, but on the more immediate concerns of this evening's agenda.

The dinner was in honor of Bernard Kouchner, the French minister of health. Kouchner, a founder of the charity Doctors Without Borders, which won the 1999 Nobel Peace Prize, was in town to speak at a meeting sponsored by the Brookings Institution. The subject of the meeting was how the United States and France could cooperate to fight AIDs in Africa. Since Haseltine was on the board of Brookings, the centrist think tank and bullpen for out-of-power Democratic Party policy wonks, and since he had been active in anti-HIV efforts since he began studying the virus back at Harvard, he was a natural host.

He also had the house for it. The Georgetown mansion was big and old and made a statement. Seeing the traditional furnishings inside, one would never guess that as a young man, Haseltine had sat in front of a police car at the start of the Berkeley free speech movement. There was hardly a telltale sign that Bill Haseltine was anybody other than a rich corporate executive with good taste in antiques and art, except, maybe, a modern-looking portrait hanging in the drawing room where guests were beginning to gather. The painting depicted Haseltine in the foreground. It was a flattering portrait, as most are, that made Haseltine slightly more dashing than he is in real life. Over Haseltine's shoulder, the artist had created an image of a marble statue of the kind carved by ancient sculptors of Roman generals and emperors.

Another portrait, a black-and-white photograph, hung in the entrance hallway. It looked like a picture that could have been taken by the mod fashion photographer played by David Hemmings in the 1966 film *Blow-Up*. It was a profile of the face of Gale Hayman, Haseltine's wife. She was once tagged as one of the best dressed women in the world by Mr. Blackwell and since leaving the glittery Los Angeles world of Giorgio's, and taking an estimated $82 million as her share of the business after it was purchased by Avon, she had started her own cosmetics line and written beauty books. But on nights like this, she also acted as Haseltine's social director.

She filled the role perfectly, greeting each arriving guest with the practiced geniality of a woman used to playing hostess to important people. There were to be about thirty at this dinner party and Gale had made sure the food would impress. The menu included rack of lamb, creamed spinach, potatoes, good wine, apple tarts for desert. There were waiters and a butler, and parking valets at the ready.

James Steinberg—no relation to Wallace—and his wife were among the first to arrive. Steinberg, the former deputy national security adviser to President Bill Clinton, was a Brookings vice-president.

Ian Wilmut soon followed. He was on the board of Haseltine's Society for Regenerative Medicine and was in D.C. for the meeting downtown. Wilmut looked like a science fair winner attending his first junior prom, a little out of place amid the arriving Washington pols. He wore a rumpled brown jacket. His unruly reddish beard wrapped his round face and his few remaining hairs crisscrossed his head. Still, nobody recognized him until Gale formally introduced him to the Steinbergs and dropped Dolly's name as a point of reference. Steinberg was instantly fascinated.

There had been a great deal of chatter about biology in the past months. Embryonic stem cells, and the debate over cloning human beings, either to make babies or to create treatments for the diseased and disabled, had been the lead story throughout the past summer. Then the country was attacked by Islamic terrorists on September 11. In the weeks after, letters laced with anthrax were mailed to media outlets and Capitol Hill. Anthony Fauci, the director of the National Institute of Allergy and Infectious Diseases, was now one of the most famous faces in America, thanks to near daily appearances at congressional hearings, press conferences, and news interviews. Tonight Fauci would headline in Haseltine's drawing room when he and Kouchner would make a few remarks about AIDs, about anthrax, about treaties that might prevent biological attacks.

So Steinberg's interest was more than polite. Between bites of smoked salmon and sips from cocktails, he and Wilmut talked over whether biology was about to change everything. Life had become so elastic, hadn't it? Was science changing society? Religion? Something profound was going on, no?

In the 1999 book Wilmut wrote with Keith Campbell and writer Colin Tudge, he declared "we can look forward to an age in which the understanding of life's mechanisms will be virtually total . . . from this understanding will come—if we choose—total control."

Total control over our biology? No wonder the book was entitled *The Second Creation*. With total control you could create anything, make any change you wanted in human beings. But while Wilmut was happy to talk about science, conversations like this made him squirm. He spends a lot of time engaged in them because he is one of Dolly's daddies and when people find out they want him to explain the meaning of it all. But Wilmut is a modest man. He started his career to breed better farm animals, not to be a philosopher. He's not sure if there is meaning, and he certainly did not want to blurt out that human beings were going to live forever or start genetically enhancing themselves. He's not sure that is ever going to happen, for one thing, and, like Haseltine, he saves such observations for ears he knows will assume the many caveats of science. Yet he did agree with Steinberg. Something profound was going on.

But exactly what the profound thing was escaped most people who were caught up in the craziness that had come to surround cloning. Just a few hours earlier, Haseltine and Wilmut had sat in the middle of the craziness with Mike West and TV cameras in a perfect example of how successful the bio-Luddites had been in using cloning to hijack the entire biotechnological agenda.

Looking around at the media scrum inside the small conference room of the Renaissance Hotel, it was hard to tell whether Haseltine had gotten just what he wanted or if he was taken by surprise. The flying elbows and the waving boom mikes and the TV cameras were not exactly like the scene outside O.J.'s Brentwood house when the white Bronco pulled up, but it was unusual for a small scientific conference like Haseltine's regenerative medicine meeting. Typically, such specialized scientific conferences draw a few print science beat reporters from, say, Associated Press or Reuters, maybe a local newspaper in the city where the meeting is held, and the science journals. The *New York Times* and the *Washington Post* might send somebody if there's potential for good gossip or an important announcement. TV never comes.

TV doesn't have the patience. A typical scientific presentation at a meeting is about forty-five minutes long. The talks almost always take place in the dark. During the first ten minutes, the scientist-presenter fumbles with a balky laptop computer in an effort to get the PowerPoint program to work. During the next thirty minutes, the scientist, who has never been trained in the art of public speaking, explains, often through a very thick Chinese, German, French, or Italian accent, why the mass of pinkish cells on the right is the surprising and highly significant result of the procedure performed on the almost identical mass of pinkish cells on the left. Line graphs are shown. The final five minutes is taken up by a question period. Colleagues stand at a microphone in the middle of the aisle and, using the polite code phrases of science, ask the presenter if he has considered the possibility that his head has unaccountably become entangled in his ass.

But this meeting was taking place in the full flush of bio-Luddism when, if somebody used the word *clone*, TV came running. Haseltine had done much more than say the word. A journal he edited had run a paper by Mike West's Advanced Cell Technology (ACT) scientists announcing they had just tried to create human embryo clones. So while the usual complement of print reporters sat around the conference room table, the TV folks surrounded it like a wall of defensive linemen. Some of the finest minds in the new field of regenerative medicine were in attendance, including Haseltine, Stanford scientist Irving Weissman, Harvard tissue engineer Anthony Atala, and Wilmut. But the man TV had come to see was Michael West.

West plopped down at the table. He looked weary. Just six days before, on Sunday, November 25, West had used the sparsely read Internet-based journal Haseltine edits, *e-Biomed: The Journal of Regenerative Medicine*, to publish the paper discussing the cloning-for-stem-cells experiment. Unusually, however, the very day *e-Biomed* was issued, *US News and World Report* trumpeted this same ACT work as "The First Human Clone" in capital letters on its cover. Also that same day, *Scientific American* posted a story about the experiment on its web site.

The bio-Luddites pounced, wielding the usual rhetorical weapons. They ignored the fact that ACT had no intention of making babies, and that the experiment had failed. There were no embryos from cloning. But George W. Bush declared ACT's research "wrong," arguing somewhat ambiguously that "we should not as a society grow life to destroy it." Congressman Christopher Smith warned, "We are on the verge of having human embryo farms in laboratories all across America." ACT's scientists were "mad doctors" and one TV commentator declared them part of a "white-lab-coated priesthood." Bioethics Project leader William Kristol worked himself into a lather. In a *Wall Street Journal* editorial, he and Eric Cohen, misreading H.G. Wells, said Stop! Before it's too late! and we wind up on the Island of Dr. Moreau with "the dehumanization of man, and the creation of a post-human world of designer babies, man-animal chimeras, and 'compassionate killing' of the disabled." As if that weren't scary enough, Kristol and Cohen evoked everybody's favorite all-purpose enemy. In Kristol's *The Weekly Standard*, they published an editorial entitled "Dr. West and Mr. bin Laden."

Haseltine suffered some serious blowback, thanks to the ACT paper, not because of the shouting of the bio-Luddites, but because within science, he was now seen as aiding and abetting West's own quest for publicity. After all, the experiment was not especially interesting and it failed, so why else would West trumpet it and Haseltine publish it? Members of *e-Biomed*'s advisory board announced their resignations. Wilmut, among others, suspected that Haseltine had arranged the publication of the ACT paper in order to stir press attendance for the meeting.

Haseltine may well have appreciated the extra media exposure for his meeting, but he also felt sandbagged. He was angry about the coordination between ACT, *US News*, and *Scientific American* and when his own motives came into question, he became angrier still. William Haseltine thinks of himself as a man of science. He scrupulously obeys the unwritten tribal code, and one of the commandments in that code is that the results of experiments must be revealed in a peer-reviewed journal or released in a scientific meeting before they are let loose on the wider public. And a scientist must never, ever, work up a cheesy research paper just to get some headlines.

In fact, ACT's paper was not cheesy. True, it wasn't the greatest in the world, but then, the journal, being brand new, was not the greatest journal in the world. The paper did what research papers do. It described the experiment, revealed results, disclosed how those results were obtained. Though Haseltine wished he had sent the paper to more reviewers, the feedback from the reviewers who had seen it was mostly positive. A few revisions were made as a result of one reviewer's objections, and then the paper was published. The fact that the experiment did not yield any embryos from cloning did not make it a lousy paper. Failures are often published in the scientific literature. In fact, many scientists think more failures ought to see print so the world of science learns from the efforts of others.

What, then, accounted for the intrascience grumbling? Through the flack of denunciation, you could see that it was not so much ACT's experiment or the paper's contents that so riled biologists, it was the timing. Haseltine had published the paper in his journal at precisely the wrong political moment. In an editorial entitled "Publish and Be Damned," the journal *Nature Biotechnology*, while not naming Haseltine, insinuated that, thanks partly to *e-Biomed*, "we can no longer hope for a sensible and thoughtful political debate of the issues. Instead we look forward to a total ban on all forms of human cloning." The House of Representatives had passed a bill six months earlier that threatened to send scientists to jail for ten years if they did what ACT had just done. The Senate had yet to vote and just three days from Haseltine's press conference, West was slated to appear before a Senate committee to testify.

ACT had been trying for several years to obtain stem cells from embryos, sometimes using cloning technology, and it was quietly assumed within the small community of cloners and embryonic stem cell researchers that others were trying to make embryos this way, too, both in the United States and in other countries. Biologists were not shocked that ACT had tried because some had done similar experiments before. But they kept their attempts quiet. So the worry was not so much "What the hell is Mike West doing?" but "Why the hell is he talking about it?" To those familiar with ACT's history, the answer was simple: West needed cash. ACT, a tiny company smaller in both physical space and number of employees than many university labs, was barely treading financial water. A flurry of press might help draw investors.

West angrily dismissed this theory, arguing that, if anything, the controversy over cloning made ACT radioactive. In an eloquent and impassioned letter to the editor in response to the *Nature Biotechnology* editorial, West, ACT's cloner Jose Cibelli, and company science honcho Robert Lanza argued they were simply acting in the name of "transparency, proactive communication, and ethical reflection." ACT wanted to shine the light of day on its experiments, even these early-stage attempts. "We do not believe that silence is a good alternative. In our opinion, the philosophy of sticking our heads in the sand, and simply hoping that the world does not pay attention to what we are doing is a blatant betrayal of the public trust."

But West also told other versions of the story. In one, the timing was the result of a snafu caused by the fact that ACT's experiments were the worst-kept secrets in biotech.

Weeks before the paper came out, the BBC had filmed Jose Cibelli, ACT's cloning specialist, retrieving eggs from a fertility clinic, entering the small room where he does his work, expressing disappointment that embryos failed to thrive. Meanwhile, *US News* had a reporter camped out in the hotel down the street from ACT's lab. The *Wall Street Journal* had discovered that ACT was using human eggs. And Lanza had agreed to help with an ACT story on this work for *Scientific American*. West was shocked, shocked, to find that all those media threads wove him into a trap.

"I asked that *US News* not learn about what, if anything, we had done with human nuclear transfer," the preferred science name for cloning technology. "Of course, if you end up pushing hard enough we'd end up saying, yeah, okay, we made some embryos." Given such strong hints, of course reporters pushed. West worked out a deal with *US News* not to run the cloning story until ACT could publish in a scientific journal. But the BBC program was set to air in January or February. "And here we would be with no paper. Then we'd get '60 Minutes' coming to our door and you'd see my hand over the lens!"

"We had planned to wait until we got stem cells and the thing worked and we could publish a big paper. But we decided, hey, lookit, this is going to take a long time and the BBC will announce it, so let's put out a paper now. We wrote it up and Bill Haseltine's journal was an Internet journal and we were on the editorial board, and he was always asking us for papers so we said, 'Let's give it to Bill.'"

West, and consequently Haseltine, had been swept up in the cloning hysteria manufactured by the bio-Luddites. But the effort to disrupt the bioutopian agenda may have backfired on everybody, including the bio-Luddites. For by making cloning their poster issue, by imbuing it with awesome, totemic power, they only made cloning that much more enticing to those who thrived on transgression, and to those who hoped it would save them from nature.

WHEN SOME OF WEST'S EMPLOYEES protested before publication that the cloning paper was premature, and that it was sure to create a backlash, he told them yet another story. He was worried that waiting would be worse than pushing ahead. Groups around the world were making, or about to make, attempts of their own to retrieve stem cells from embryos made by cloning, he said. But he was especially worried about the freelance cloners like Panos Zavos and his one-time partner in cloning, Italian IVF doctor Severino Antinori. They had announced back in February that they would attempt to make actual crying, squirming babies by cloning and what if they came up with one—or even a pregnant woman—before ACT could show the therapeutic value of cloned stem cells? And if the freelancers were not scary enough, imagine the international backlash from the worst case of all: a Raelian clone. Yes, the cloning hysteria had become so bizarre that West told his employees he was frightened that Rael—a French amateur race car driver turned UFO sect leader, a man who tied his hair in a funky top knot, dressed in a trademark white spaceman tunic, his galaxy medallion draped over his chest, a man who handed out business cards listing his professions as "race car driver," "Founder of UFOLand," "Founder of Clonaid," "Founder of the Raelian religion," "Best-selling author," and "author-singer-songwriter"—might stand in front of a CNN camera kissing the first baby clone.

Hardly anybody who had looked into the Raelian cult's claims to be actively trying to clone people thought it had a chance of doing anything other than flubbing the job. After all, Rael had been promising a clone on the ground—any day now—for four years. But these people made such a weird story that they received more free TV time than the Dallas Cowboy cheerleaders.

West told his employees that he wanted to show something, anything, that could demonstrate the potential of nuclear transfer as a therapeutic technology before Rael could push freaked-out politicians into legislating a total ban. He was not wrong to worry. A ban had been the first response from lawmakers when Panos Zavos had said the magic word. There Zavos was, passing his days in Kentucky collecting and analyzing sperm, a man who knew almost nothing about cloning. But then, in late January 2001, *Wired* magazine declared that human cloning might happen soon and two weeks later Zavos made his February announcement that he'd try to clone a person. The next thing he knew, he was testifying before a U.S. Congress subcommittee on March 28. He was so uninformed about the science of cloning, in fact, that much of his testimony had been written for him by a mystery man, supposedly a British scientist, calling himself "Roger Moorgate"—a key member of the international "cloning movement." Zavos became an instant celebrity who demanded $2,000 per interview from reporters, changed the name of his company to the Zavos Organization, and was reportedly working on a movie deal. All this before he had conducted a single cloning-related experiment.

That was just the sort of hype the Raelians tapped into from the first days of Dolly. To them, Dolly was, literally, prophecy come to life.

The Raelians were just another obscure UFO cult known for attracting nerdy-looking men and cute women who sought wisdom through "sensual meditation," a practice that necessitated a warning to Raelian training camp attendees to "have a medical checkup before you come to the camp to check for transmittable diseases, particularly of a sexual nature. The use of condoms and an understanding of safe sex is essential." But when the Roslin team that created Dolly published news of her birth in the February 27, 1997 issue of *Nature*, they were able to argue that their own rapture was around the corner.

Thirty years ago, on December 13, 1973, Claude Vorilhon saw a UFO land in the French countryside. The UFO looked exactly like the one that disgorged Michael Rennie in *The Day the Earth Stood Still*, so, naturally, Claude approached. He met a spaceman. Claude and the spaceman got along well, meeting one hour each day for the next six days, in a series of instructional seminars on the flabbergasting news that God had nothing to do with life on Earth. Rather, Earth had been seeded by Elohim, extraterrestrials who had perfect command of genetic engineering and DNA synthesis. (The fact that the first gene splicing experiments had recently made worldwide news, was, apparently, just a coincidence.)

Unfortunately, the ancient human beings created by the Elohim mistook their creators for gods. This led to the development of a number of false religions. But the Elohim would not be deterred. They sent a series of prophets to shepherd humanity toward the bright future that was waiting at the end of the technology rainbow. Once human science had advanced far enough, humans could rejoin the Elohim. That time was almost here. Claude was told that he, freelance auto racing journalist, aspiring amateur driver, was to become the last in a line that had included Moses and Jesus and Buddha. The Elohim dubbed him Rael, "messenger."

Two years later, on October 7, 1975, Rael was taken to the home world of the Elohim, a planet located one light year from Earth. (The Elohim maintain a forward base to keep an eye on Earth. The base is located in the hollowed-out center of the moon.) There, it was revealed to him that Jesus had indeed risen from the dead, but not through any miracle. The Elohim had saved one of Jesus' cells before the crucifixion. The risen Jesus was a clone. Rael returned to Earth and predicted that cloning technology would come to humanity in the final step of the long march toward reunification with the Elohim and immortality. (The fact that cloning was much on the minds of some people in the run-up to the first IVF baby in 1978 was, apparently, also a coincidence.)

Dolly was a sign of the Second Coming. "We were all so happy!" one Raelian told me. "All over the planet we were sending e-mails to each other after Dolly." Rael chose the Flamingo Hilton in Las Vegas to proclaim more good news. On March 11, 1997, just three weeks after the news of Dolly had been published, Rael announced that he and unnamed investors had formed a new company called Valiant Venture Ltd. Valiant Venture would consist of three subsidiaries: Clonaid, Clonapet, and Insuraclone. Clonaid would clone people, living or dead, for the small fee of $200,000. Clonapet would assure families that Sparky could join them in everlasting life. Insuraclone would store the cells of the living so that, should you be run over by an International Harvester while crossing the street, you could be resurrected just as Jesus had been.

Brigette Boisellier, a chemist who worked for the French chemical giant Aire Liquid, was named scientific director. She explained that labs were being established in the Bahamas. The site may have been chosen by Rael, who, after being hounded out of France, had found a congenial home in Quebec but had come to dislike the cold weather.

"We are so eager about cloning," Boisellier said. "Rael wrote that one day we would be able to be cloned and one day we will be able to transfer the personality into the new body so that there would be eternal life. He wrote that twenty-seven years ago. Imagine all the scientists saying, 'Come on, you know this is impossible. The DNA code cannot be reactivated.' But I trusted Rael and believed in what he wrote." And then came Dolly and the word became flesh.

But while Dolly was real enough, the Bahamian labs were not. There never were any Bahamian labs. No experiments were ever done. In all the years since the creation of Clonaid, there has never been the slightest shred of evidence that Boisellier, Rael, or any scientist affiliated with them was seriously attempting to make babies by cloning. This did not stop them from attracting customers, however. They did collect money from at least one couple from West Virginia, Mark and Tracy Hunt, who ponied up a reported half million dollars to Clonaid. Boisellier set up a "lab" in Nitro, West Virginia, in a former high school classroom converted to business use that was so woefully equipped the success of any biology experiment would have been a miracle. Mark Hunt soon fired Boisellier, calling her "a press hog," an apt description borne out when Boisellier announced in late 2002 that the Raelians had made half a dozen women pregnant with clones and that one had given birth to the first cloned human, named, naturally, Eve. Again, there was no evidence, and besides, the success rate claimed by Boisellier would have instantly made the Raelians not only the best cloners in the world, but the best IVF doctors, too.

Boisellier was perfectly willing to make things up on the fly, even how cloning fit in with her religion. Raelians believed that the soul resided in the DNA and that DNA emitted signals that could be received by the Elohim. "To the Elohim we have personal identity based on the cellular code." But what about identical twins who have the same genetic code? How do the Elohim tell them apart? Wouldn't cloning, which duplicates the genetic code, garble our own personal messages? Boisellier was flummoxed, finally suggesting that "there are some cells where there is more information than just DNA code and they are different from the one when you are born. They are probably around here," she said indicating her forehead just above the bridge of her nose. "That is something we will have to check . . . There are some things in the Book of Rael we will have to change. One day we will have to research that, but don't mention that to Rael."

But just as it had with Zavos, Congress lapped up Rael. It handed Rael and Boisellier the single biggest dose of legitimacy and PR they had ever had when it called them to testify at the same hearing as Zavos.

Yet while all indications were that the Raelians were faking it, the "cloning movement" was actually gathering steam. Though it was hardly the worrisome power that some bio-Luddites made it out to be, it would, eventually, spawn a bizarre underground race to clone a human being.

RANDOLFE ("RANDY") WICKER knew better than anybody that the one sure way to get yourself into the newspaper or on TV was to say you favored something everybody else condemned. Wicker, born Charles Gervin Hayden Jr. in 1938, first tapped into his own lust for publicity while a student at the University of Texas. In 1960 the campus *Texas Ranger* magazine profiled Wicker under the title "Arise!! The Thrilling Story of One Man's Fight against the Forces of Virtue, Peace, and Complacency." Half a dozen photos showed a thin, rangy, jacketed-and-tied Wicker in various stages of froth, haranguing a crowd, collecting signatures for petitions. "To some he is a comic mascot," writer Ray Hanson stated. "To others a sharp-fanged, slavering villain." For example, Wicker demanded that blue nose citizens stop removing girly magazines from Austin's newsstands and, thanks to his efforts, the racks were made safe for June Wilkinson once again.

Girly magazines were one thing. Open homosexuality was quite another. Unbeknownst to his family, Wicker had joined the New York chapter of the Mattachine Society, one of the earliest "homophile" organizations in the world. Though Wicker attended meetings when he was home in New York, his father did not discover his son's secret until he read the diary of Wicker's freshman year at a Virginia college. Later, Wicker would insist that his father handled the situation about as well as a father could in 1958. He just asked that Wicker never allow the name of Charles Gervin Hayden to be linked to homosexuality. Wicker agreed, making up a nom de sexualite, spelling Randolfe with an "fe" so that, "should my name ever appear in lights," nobody would ever forget it. He chose Wicker because it sounded a little like "wicked." He eventually made this his legal name.

The Mattachine Society came to regret admitting its youngest member. The society had survived by spreading news of meetings through word of mouth. Members kept a low profile. But Wicker was all about high profile. "I had gotten so frustrated with the gays," Wicker remembered. "I figured they would never come out of the bars." On July 15, 1962, radio station WBAI conducted the first ever broadcast interview of a group of homosexual men in New York. Wicker had arranged it. When columnist Jack O'Brian of the *New York Journal American* heard about the station's plans, he condemned WBAI and called Wicker "an arrogant, card-carrying swish." Wicker immediately began advertising himself as an arrogant, card-carrying swish.

Some members of the community came to the conclusion that Wicker was more trouble than he was worth, a crank and a stunt artist. But a few younger members admired him. "Randy had achieved a kind of mini-fame," gay journalist Jack Nichols wrote. "He was a gay media star, one who boldly used his legal name!" O'Brian was right about one thing. Wicker was card-carrying. He had business cards printed up that called him "public relations director" of the "Homosexual League of New York." He was also the only member.

Wicker was out and he didn't care. Being out was good. Being out, and being vocal about being out, got you a lot of press in the early 1960s. He was profiled by writer Dick Schaap. He made more radio appearances and got his name in the mainstream newspapers by picketing a Manhattan U.S. Army induction center with a few other early gay rights activists.

By the mid-1960s, Wicker was an official member of the Greenwich Village scene. If it was counterculture, if it was outrageous, Wicker tried to be part of it. He helped poet Allen Ginsburg launched LEMAR, the League for the Legalization of Marijuana. He bought beads from Abbie Hoffman. Wicker and a lover opened a psychedelic shop in St. Mark's Place called Underground Uplift Unlimited and Wicker began creating buttons with pithy, provocative slogans to sell in the shop. The first, "Legalize Pot," was a smash. Other big sellers included "Lindsay Turns Me On," referring to New York mayor John Lindsay, and "F*CK Censorship." By 1967 and the Summer of Love, Wicker had sold two million buttons and was known as the button king of New York.

But once society caught up to Wicker, he was no longer a star. In 1969, when the New York City police rousted a gay bar called the Stonewall Inn, the drag queens who milled about outside revolted. One, Sylvia (born Ray) Rivera, a seventeen-year-old Puerto Rican teenager and sometime street hustler, was among the first to throw a bottle. The ensuing riot is now considered the Bunker Hill of the gay rights movement, but Wicker spoke out against it, arguing that "rocks through windows don't open doors." At subsequent meetings of activists, Wicker verbally battled Rivera. But Rivera had become a hero and in this new atmosphere, Wicker was quickly making himself irrelevant, a bit like the old-timers of the Mattachine Society.

Wicker eventually settled into domesticity with a lover named David Combs. The couple opened an antique lighting store, Uplift, on Hudson Street and moved across the river to Hoboken. Then AIDS took even that life away from him. "Nearly half of those receiving this letter in years past are either dead, or infected and waiting," he wrote to friends in a 1990 holiday letter. "If this holocaust has spared you and yours, do rejoice and let me rejoice with you. . . . I've run out of tears. The aches inside have turned to numbness."

Since the day his own father asked Wicker to forsake his family name, Wicker had gathered ad hoc families around himself, and now most were gone. He was lonely. Gay neighborhoods were being gentrified, tourists came to Greenwich Village to titter at the leather bondage gear in shop windows, mayors made a point of walking in gay pride parades. An outspoken gay man was just another colorful part of the city. There was nothing for Wicker to do but sell lamps, and try to reassemble a family. He took in drag queen crack addicts, homeless transgendered, teenage runaways. He patched things up with Sylvia Rivera, who had become a homeless drug addict living on a pier, and gave Rivera a job in the store. David's ashes sat in a box above the cash register.

Then Dolly came into his life and Wicker, who knew nothing about biology and did not care to, reacted to the reaction. Just eight days after the *Nature* paper was published, a New York Republican state senator named John Marchi introduced a bill that banned human cloning and mandated prison for anybody who tried it. "We

ought not permit a cottage industry in the God business," Marchi
said. Wicker knew an ideologue when he saw one. "I saw these were
the same social conservative, antigay people coming out against
cloning. I thought, 'Boy, this is weird. Why are they so upset?' I
thought it over and realized hey, three females. One gave the cell,
one gave the egg, and the other bore the sheep. Same sex reproduction! That's it! They understand you no longer have to have male and
female to have offspring. That got me to the idea that these were the
same homophobic ones."

Biotechnology and gay rights merged in Wicker's head and he
saw a chance to make his star shine again. Wicker press-ganged
Sylvia, some friends from the neighborhood, and a few homeless
people who knew Wicker was a soft touch for money to distribute
flyers and mount a protest. He went to city hall and registered the
name of his new group, Clone Rights United Front (CRUF). Once
again, he was the only member.

At first, he'd say anything provocative to get the cloning story
some play, even mounting a "clone Diana" demonstration in Central
Park during the city's memorial service for the princess of Wales. But
then Wicker bought a computer and through the magic of the Internet, came into contact with Mark Eibert, a San Francisco Bay area
attorney who was involved in infertility issues there. Eibert saw
cloning not as a gay rights issue, but as a way for profoundly infertile
couples to have genetically related children. The two began an e-
mail and phone correspondence. Other Internet contacts developed,
Wicker's own views became more sophisticated, and by 1999 he had
become a board member of the Human Cloning Foundation, a
501C3 charity registered in Georgia. The other board members
included University of Texas lawyer John Robertson, the legal counsel for the American Society of Reproductive Medicine (the IVF doctors); Gregory Pence, a University of Alabama Birmingham philosopher and bioethicist; Nick Hennenfent, an Illinois doctor; and his
brother Bradley, a doctor in Atlanta. Best of all for Wicker, because
none of these men wanted to be identified as pushing cloning, they
made him chief executive. "I am the CEO!" he bragged. "I am going
on Fox news and getting to be the mouth that talks back to whatever
idiot they drag up to go against me! I am bigger this way, not just the

Wizard of Oz behind the CRUF, the man behind the curtain with the big microphone with no troops to speak of."

As a player in "the movement," Wicker also managed to get himself time before Congress to testify during the original round of post–Dolly cloning hearings, and, like the cryonicists, began casting a net for anybody famous who wanted to be cloned. On June 26, 1999, he wrote a letter to physicist Stephen Hawking. "Some of your admirers, with the best of intentions, have suggested that you should be asked to (or allowed and urged to) have samples of your DNA frozen so that your genotype can live on into another lifetime . . . Please hear the pleas of those asking that samples of your DNA be frozen now! . . . I ask that you DEMAND some of your DNA be used in cloning as soon as safely possible."

Hawking replied through his assistant: "Professor Hawking thanks you for your letter dated 26 June, 1999. He thanks you for your offer of cloning. He prefers to reproduce in the traditional way."

By mid-2000, Wicker had actually managed to attract a lot of traffic to the foundation web site. Many correspondents were, like him, defiant contrarians. Many were cranks. Some were clearly unbalanced. Some were like the Hunts, desperate souls who had lost children and were operating under the delusion—fed by society's own sci-fi fantasies—that cloning would bring those people back. A few were infertile couples who had tried and failed at IVF. The mysterious "Roger Moorgate" signed onto the site and began taking an active role, supplying very real science information, helping prepare Zavos's testimony, answering questions for those who queried. Later, when the Human Cloning Foundation fell on hard times, and the Hennenfents' identity was exposed, they withdrew. "Roger Moorgate," Zavos's ghost writer, stepped into the breach. He and Wicker collaborated on a new web site called Reproductivecloning.net and Roger Moorgate went public. Amazingly, it turned out he was actually James Byrne, a young protégé of Cambridge University scientist and cloning pioneer John Gurdon.

Wicker carved a claustrophobia-inducing space out of the clutter of his store, equipped it with a phone and a computer, and decorated it with a poster-size photo of himself as a little boy—his later-born twin, he would say. Sylvia sold lamps. A drag queen and struggling crack addict named Coco ran errands and did odd jobs to pay off Wicker for allowing him to sleep in the back. Homeless people wandered in and hit Wicker up for money. Sometimes they would walk in with lamps, obtained from who-knows-where, and ask Wicker to buy them. The last thing Wicker needed was another lamp. But he always paid. When evening fell, Sylvia's lover, Julia, would arrive and everybody would trade barbed jokes. There Wicker would be, with his gray hair, his glasses sliding down his nose, a little perspiration showing through his business shirts with the button-down collars, in a gab fest with Sylvia, a transvestite; Julia, a partial transgender; and Coco. "Isn't it wild?" he once said. "I'm the most normal one here!" The store would shut down, then Wicker would pour himself a giant plastic cup full of vodka and Squirt, climb into his cubby hole, and work the computer for hours. This was the nerve center of the world's human cloning movement.

Aside from Byrne's participation, it was entertainingly absurd, more a window onto just how willing many people were to use biotechnology as a tool to achieve happiness in defiance of wide-spread condemnation than it was a real movement. But as was true of the early Bay Area computer scene, the Extropians, the cryoni-cists, and even Wally Steinberg, people imagined the rapture first and then tried to make it come true. Everything the cloning wanna-bes had heard from the bio-Luddites and from the great, anxious majority only convinced them that cloning would work and that it could provide an out for whatever emotional jail they were in.

One scientist who communicated regularly with Wicker wrapped a philosophy around cloning, sure that it would provide a quasi-immortality for potential customers and everlasting fame for him. "This is the next best thing to eternal life," he insisted, because cloning was a form of resurrection. Life was all about genes, he said, and cloning would let your genes live on. "We must remember that the desire to be immortal even only a second best immortality (that is cloning) is the STRONGEST in a human being. Human beings will do ANYTHING to be immortal. I KNOW this."

Though it was obvious that the scientist was a loose cannon, he did manage to attract several people, "clients," who were serious about wanting him to clone. Negotiations with one client eventually broke down over money and whether the scientist could guarantee success in cloning the man's dead son. But he kept trying, constantly fishing for new "clients." He had read that a man named John Sperling, a wealthy eighty-two-year-old businessman, was the money behind the cloning of a cat, Copy Cat, and wondered if he could entice Sperling. If not, how about Marvin Minksy, the MIT artificial intelligence guru? He had signed up to be chilled by Alcor, hadn't he? Or Max More, a leader of the Extropians? The Extropians seemed like smart, open-minded people. They were true believers in the power of cloning, weren't they?

In 2002, he claimed to have launched a failed attempt to clone his own girlfriend, a former cocktail waitress-turned-biology-student, in a New Delhi IVF clinic. "I feel like Christopher Columbus," the girlfriend said. Cloning fear was just as irrational as thinking you'd fall off the planet if you sailed west. "People just don't understand that this is the next stage of our evolution."

At one point, the scientist offered to clone Wicker. "I am especially compromised since I have been telling media that 'independent operators have offered to clone me at minimal expense and I have declined their offers,'" Wicker said.

"You have to understand that I WAS the 'homosexual civil rights movement' in the early 1960s but when the Stonewall riots occurred in 1969, I was 'too conservative' and urged people not to riot—that 'bricks through windows do not open doors,' etc. That was, to date, the greatest mistake of my life. Now, once again, I feel like I stand at the portals of history. Will I, once again, be a 'numbnut' who is afraid to 'go for the prize'? Will my conservatism, once again, cause me to be just an 'echo' in history?"

But Wicker knew that the focus of the press had moved on. At one time he was the star, but now reporters just pumped for information about others. He had become the easy girl under the bleachers who realized the shallow intentions of suitors. "Frankly, I have gotten tired of the stupid media. This media whore has gotten tired of turning tricks only to have them shit all over me."

Sylvia Rivera died of liver cancer on February 19, 2002. After Rivera's funeral, Wicker had gone to the Metropolitan Community Church, where Rivera had begun working as a coordinator for a food bank. The people at the church told Wicker that she had spoken fondly of him. "Randy had always been so good to me," Rivera had said. When Wicker recalled Rivera's words, he began to cry. Another family member gone.

A short time later Wicker appeared on Comedy Central's "Daily Show with Jon Stewart," in a goof about cloning. Wicker wore a Stars and Stripes shirt, and tossed a baseball back and forth with himself.

THOUGH THE WORLD was certainly diverted by cloning and the comic international detective story of who might be cloning whom, the smart money never bet on human cloning at all. The real point of Dolly was not that you could make a sheep. The point of Dolly was that a cell did not have to be what it was. That was the profound thing Wilmut knew was happening, the thing that made James Steinberg so curious. Dolly was walking, eating, baah-ing proof that Haseltine's notion of "transubstantiation" was real. Biotechnology could turn water into wine.

The mammary cell that made Dolly was not a mammary cell anymore. A cell that was at a dead end, with no place to go but cell heaven, got a do-over and became liver cells and heart cells and skin cells and nerve cells. Biologists knew what this meant, especially John Gurdon, who had been saying it for forty years. "I personally believe that in time we will create any cell we want and any organ we want. We know that this must be possible." Why? "Since we now accept that . . . you can switch cell function dramatically, this opens up the possibility of making any cell from any other cell." Bioutopians and transhumanists got it, too, and they realized that everything they had ever hoped for was becoming part of mainstream science. As one character explains to Heinlein's Lazarus Long in 1973's *Time Enough for Love*, "every cell in the body contains a full bundle of chromosomes. I suppose that you could grow eyes, or bones, or anything you liked anywhere, if you knew how to manipulate the genes . . ."

"If there is a path to human immortality that is it," Haseltine said. "First it will be for health and regenerating tissues, but then it will be for rejuvenation." The trailhead to that path, he said, was the ability to reprogram cells. "Nobody believed Gurdon. They didn't think it could be done. [They said] that it was too complicated. That was a big mistake."

Tanja Dominko believed Gurdon. She thought cloning people was a silly idea and even though she was the editor of a journal entitled *Cloning*, she wasn't all that enthused about cloning animals except as science experiments or to make special agricultural animals. "Cloning is a magic trick," she said. "We have no idea how it works." She had something else in mind, "transdifferentiation," taking one cell and turning it into something else without cloning, without eggs, without ES cells.

When Dominko first thought of this, she was almost alone, though others would soon catch up. Like most other scientists who came out of the world of agriculture, she was not afraid of radical ideas because she herself had come out of left field. She was born and raised in Slovenia, graduated from vet school there. In fact, though she came to the United States at twenty-eight to attend graduate school at the University of Wisconsin, had then gone to work at a research center in Oregon, and moved to Massachusetts to work for Mike West at ACT, she still sounded as if she had just walked out of a dark Ljubluana café. The effect was exaggerated by her good looks, her habit of smoking one cigarette after another, and the way she would argue over the color of the sky if you gave her half a chance. She was a serious scientist but she did not feel bound by many of the conventions of science or the behavior of those reared in the stringent style of the elite tribe. She was keenly aware of being a woman who was first trained as a vet in another country so she had to battle for acceptance into the fraternity. She succeeded by thumbing her nose at it. She preferred the blues to Chopin. She could swear like a sailor in three languages and slam back a beer like a longshoreman. She'd been known to wear miniskirts during her conference lectures so, she said, the old men would pay attention. The nameplate on her desk read

Tanja Dominko
Science Babe

Dominko had a dark view of the world. Her father, an airline pilot, died when she was a teenager. He and Tanja were climbing in the Slovenian alps, miles away from help, when he collapsed with a heart attack. She cradled his head in her arms as he died. Perhaps it was this experience, perhaps just her native soil, that gave her an eastern European fatalism, illustrated by her love of cigarettes that continued unabated even after a bout with a brain tumor. She shared none of the transhumanist dream. "Life sucks" was a personal motto.

Yet she was constantly amazed by biology and was an aficionado of the "old science," as she called it. She spent hours traveling through the time tunnel of libraries looking at old copies of science journals. She'd have an idea, some new experiment she wanted to try, and then she'd look, and find out that something like it had been tried in the 1950s or the 1960s. She sometimes lamented that everything had already been done.

This was not true, of course. Not literally. But it did occur to Dominko that many of the answers she sought, and the answers Mike West and the transhumanists and the life extensionists sought, were to be found back in the old journals when science was not quite as specialized and as planned to the minute detail as it was now, when biologists were open to simply wondering about the crazy things nature did, like let lizards regrow tails and planaria grow entirely new bodies.

Human regeneration existed, too. The human liver, as Gurdon liked to point out in his lectures, renewed itself in a dramatic way. Transplant surgeons were able to slice off a chunk of liver from one person, install that piece in another, and both patients would grow complete livers. The Greeks recognized this. Prometheus was chained to the rock as punishment for giving man the gift of fire. Every day an eagle would torment Prometheus by eating his liver, but Prometheus would not die. His liver would regrow during the night and the torture would start again in the morning.

The only way this could happen, Dominko realized, was if some cells around the spot of the injury could switch their jobs, go backward into some lesser differentiated state, then go forward again to grow new parts. She did not see any reason in particular why such a thing would be impossible for humans. As an ag scientist she was well aware of Gurdon's work, she had worked in the Wisconsin lab that came close to being the first to clone a mammal, she followed the experiments of everybody in the field. She also read up on experiments done in the 1980s by Helen Blau, a Stanford scientist who fused two different cell types together and succeeded in changing the cells' jobs. Dominko was among those who knew that biology was astonishingly flexible, more so than most had ever guessed.

·

So Dominko decided there had to be a way to take cells that did one thing, and turn them into completely different cells that did another on demand and without cloning, without making embryos. First, she cultured a type of skin cell in a dish. And then she tried a variety of substances she reasoned ought to transform the cells. One morning, when she came back to work and checked on her cells, she discovered she had a dish full of nerves. Apparently, the skin cells had everything they needed to be nerves, they just weren't. They were skin. But they could be nerves if ordered to be. Water into wine.

Whenever she thought about the miracle substance she used to "transdifferentiate" the skin cells, she laughed. It just sounded so ridiculous. It was an extract of frog eggs. "I don't think anything is impossible in biology anymore," Dominko said. "Not after what I've seen happen."

By 2001, transdifferentiation was hot. One company was trying to use it to grow hair on bald men. Others, like PPL, which was inspired by Dominko, used extracts from other egg types. Geron was trying to use extracts from ES cells. Not everyone in biology was convinced that transdifferentiation was real, but a lot of people were suddenly turning cells into all kinds of things.

If you really could mine your own body to turn cells into whatever you happened to need, why not body couture? A few transhumanists were already making lists of human add-ons, like a third eye on the back of the head or new sex organs. Maybe skin could change with mood. "Once you can start with a population of [skin cells] there is nothing you cannot do before you change them or after you change them," Dominko said. "Gene targeting, knockouts, replacements, you name it. Mutations, change the genes, splicing. Nothing you can't do anymore. It's all possible. Absolutely."

Haseltine thought so, too. He outlined just how profound a shift was occurring the day of his dinner party when he called the second annual meeting to order in 2001 in a small theater of the hotel. He stood at the podium, his thin hair slicked against his skull in his trademark style, his slightly rheumy early-morning eyes peering from owlish glasses, and delivered a Heinlein vision. "At its most simple, regenerative medicine is the ability to replace and restore normal health, organs, tissues, and cells damaged by disease, injured by trauma, or worn by time. That is a very high aspiration but it is one all of you know is coming into clear focus."

New drugs were being discovered based on recipes found in our genes. Cell technologies meant that organs could be re-created in laboratories. Nanotechnology would harness the smallest bits of matter so we could become a mix of organic and synthetic. All of it, everything, was going to merge to create the future of man's physical survival.

Haseltine claimed to have invented the term *regenerative medicine* in 1999 when he began his public advocacy for it. He took great pleasure in seeing it become a slogan and then spread all over the world. Often, at scientific or biotech industry conferences, some speaker would have a Haseltine quote about regenerative medicine embedded in his PowerPoint presentation. Haseltine, however, did not invent it. The concept of regenerative medicine was an old one and the term itself was used as early as 1992 when a Maryland company called Osiris began using the phrase in its corporate literature. Whether he invented it or not, though, Haseltine successfully made it his own by becoming its most ardent, and prominent, supporter, seeing its potential early and then using his considerable weight to push for the concept before it was widely accepted.

Haseltine laid out an expanded version of regenerative medicine by nailing a collection of Lutherian theses to the door of biology. The events of the late 1990s—Dolly, stem cells, genes—had convinced him that his destiny was nigh. "We are embarking on a great enterprise: the formation of a new field of medicine that will improve the lives of people everywhere," he announced. "The key insight of regenerative medicine is that every human being was once a single cell with the potential to transform into an adult body. Each of our cells retains that remarkable potential in a latent form. We have over the past decade learned how to identify the molecules that our bodies use to direct that great unfolding. We can now isolate, study, and produce those substances in virtually unlimited quantities and use them to regenerate our tissues and organs."

As proof of what his own company could do to contribute to the cause, Haseltine showed a time-lapse video of its lead drug candidate, a human protein called KGF-2, known by its trade name, Repifermin. The video depicted Repifermin being applied to an open wound that had stubbornly refused to heal. The wound closed in stutter-step stages until it was closed completely and hardly any sign of it remained, as if by the hands of Jesus.

Drugs like Repifermin would be just a first step. Once the cells and the genes that controlled them could be manipulated, human beings would be like the 1956 Chevys still cruising along Havana's Malecon. Biologists would become like the Cuban mechanics who fashion new carburetors out of Pepsi cans and restore the pre-embargo cars long after they should have died. Better yet, humans could be souped up. Cosmetic surgery would not use implants, it would use a person's own tissues. We could be enhanced and our enhancements would be real because they would be us. No "rock hard boob syndrome," Haseltine said. A woman could have "nice, soft, fleshy boobs." The sky was the limit.

With all this new science rising just above the horizon, it was only natural that a few in mainstream science and a few bioutopians overcame their skittishness and found their way into each other's arms. "There is a merging going on," said West, who has spent his career straddling both worlds, although "the two sides are not hold-ing hands yet." Yes, but they were becoming increasingly chummy. One month after the first meeting of the Society for Regenerative Medicine in 2000, Durk Pearson and Sandy Shaw's web site, Life Extension News, trumpeted a Haseltine quote: "The real goal is to keep people alive forever."

Haseltine was famous for such statements, but he was only being more public and speaking more plainly than other scientists who had come to some of the same conclusions but who refrained from such talk even if they believed it. Haseltine knew that when some people heard him talk like this, they thought he was a nut. But nutty ideas, sometimes inspired by the scientific or cultural fringe, were begin-ning to seep into biotechnology. They were ideas that had cropped up in conversations over drinks in bars as scientists, at first stunned by the results of experiments they were carrying out, began to whis-per about the possible implications. Now, just a few years later, com-panies were being founded on some of these very same nutty ideas. In a big feedback loop, the fantastic was leaking back out of the biotech world to inspire those who were now surprised to find that their sci-fi notions were becoming real, not just the idea of remaking the human body, but even letting it go on and on, way past 100. After all these years of struggling to get real science to pay attention, the mainstream had at last discovered life extension.

> *D. E. Shaw & Co., a leading investment and technology develop-*
> *ment firm with majority ownership positions in several drug discovery*
> *ventures, is seeking an expert in the biology of aging to identify and*
> *assess long-term opportunities for pharmaceutical interventions hav-*
> *ing the potential to attenuate the aging process.*
>
> *We are particularly interested in exploring the long-term poten-*
> *tial for developing drugs that target the molecular mechanisms of*
> *aging in a manner analogous to caloric restriction . . .*

D. E. Shaw, a highly respected securities trading firm with a private equity investment operation, had identified antiaging as an emerging "space," hip business lingo for a field of business play. The idea of aging as a space would have seemed laughably silly just three years before. But mainstream venture capital and mainstream science had come to believe in life extension. The Shaw ad was not even the first sign. Shaw was merely following what was becoming a trend in biotech to convert the news coming out of labs into new businesses. Indeed, biologists involved with the new biotech start-ups in the aging space were beginning to sound even more optimistic about the potential to slow or even halt the aging process than people who were once considered crazy fringe.

At least three companies had already been formed before the Shaw ad appeared. They were created very much in the Wally Steinberg mold, the plucking of scientists from powerful institutions, the recruiting of a businessperson to run it, and the seed money all wrapped around a "story."

The story had a lot to do with C. *elegans*, a very unassuming worm, about a millimeter long, that lives in soil. It spends its very short life, typically about twenty days, vacuuming up bacteria and making baby worms. The great thing about C. *elegans*, though, is that you can see through it. In fact, under a microscope, you can see individual cells. This made it an ideal candidate for the study of development, how the genes direct cells to make a body. A scientist named Sydney Brenner, a protégé of some of the early physicists-turned-biologists and a founding father of molecular biology, thought C. *elegans* would make an ideal step ladder to boost biology up from E. *coli* bacteria and phages into more complex animals. When Brenner landed at Cambridge University, he adopted C. *elegans* in his lab. Eventually, work in Brenner's lab by another biologist named John Sulston mapped the developmental fate of every single cell in the C. *elegans* body, all 959 of them. (Brenner and Sulston were awarded the Nobel Prize for their worm studies in 2002.) For the first time, scientists could witness the complete unfolding of a creature from start to finish.

In the decades that followed Brenner's adoption of the worm as a model, C. *elegans* became a universal lab animal, joining E. *coli*, fruit flies, yeast, mice, and monkeys. Many of the "worm people," as a subset of biologists came to be known (as opposed to "fly people," say), either trained in Brenner's lab or trained under biologists who trained in Brenner's lab.

There are hundreds of thousands of C. *elegans* at the Buck Institute for Aging Research, an institution that sits high atop a hill in Marin County, California. The institute was designed by I. M. Pei and displays his love of triangles and pyramids. Because the builders used white marble from the same quarry as did the builders of the Getty Museum in Los Angeles, when the sun hits the building, it gleams like a bright, white beacon, a shiny modernist temple.

The Buck spent years existing as an idea, as an imaginary science cloister where researchers could work on one of biology's knottiest problems unharnessed from the duties of teaching university students or managing academic departments. Finally, after a local battle over the future of the Buck Trust, a financial legacy left to Marin County by Leonard and Beryl Buck, who made a fortune importing fruit and drilling for oil, a portion of the trust was granted to the institute. Assured of a constant flow of money, the institute's backers started construction and the labs finally opened in 1999.

When it opened, the Buck Institute immediately became the largest freestanding aging research laboratory in the United States. But like every other institution or person who tiptoes into the field of aging, it did not want to sound crazy, so it explained to anybody who would listen, especially concerned local citizens, that its goal was to increase "the healthy years of life." It wanted to delay or prevent the dreaded diseases of aging like cancer, Alzheimer's, Lou Gehrig's, Parkinson's, heart disease. Enhancing the quality of life for aged people was its noble purpose. The institute was not dissembling. It really did want to achieve these stated aims, but what scientists at the Buck were discovering was that the mechanisms to prevent the litany of diseases were also the ones that increased life span itself.

All I had to do to see the truth of this was look through a microscope. A worm filled the view, cruising through the never-never-land of his clear, gelatinous territory. He wriggled over the gunk, leaving a trail behind, lifting his head to sniff for something to eat, trekking onward, a brave little sod buster in a petri dish prairie. (Anthropomorphizing is a form of secret handshake to worm people, a sign that you're one of them, or that you at least *get* them, though it may not technically be a he, or even a she. Some *C. elegans* are hermaphrodites, a convenient evolutionary trick which means that, in a pinch, you can have sex with yourself.)

Though they refer to their worms in the most human terms, worm people are unable to become too familiar with individual ones, and not just because they spend a lot of time dicing them up into paste. Each worm can spawn 300 or so young ones and once they do, it is virtually impossible to tell one worm from another. When Nicole Jenkins, a young Australian Ph.D. whose worm this was, first scanned the dish looking for animals she wasn't sure how many times she had seen the same one. She finally decided to show off this one, an especially fine example of wormhood.

His handsomeness was this worm's most interesting characteristic. By rights, he should have been dead, or almost dead. He was the same age as worms Jenkins displayed under the scope just a minute before, but he looked nothing like those other worms. They were tattered around the edges, pitted with holes, barely moving. This guy was robust, a broad-shouldered he-worm.

Jenkins's boss, Gordon Lithgow, had said that "science fact and science fiction are so close right now. We squirt something on worms that makes them live longer. I mean, that's science fiction!" Clearly, science fiction had happened to this worm. He was born just like all the other worms, but then he was visited by the lab-coated god who dropped a chemical solution onto him and was given the gift of long worm life.

Lithgow did not start his career with a view to studying aging. He was raised in a gritty industrial town just outside Glasgow, Scotland, and though he was interested in biology from a young age, as a middling student with mild dyslexia, he assumed the best he could do, if he worked hard, was slake his interest by studying microbiology at a local technical college. With a certificate in hand, he might wrangle a job in a brewery fine tuning yeast or in a whiskey distillery. But Lithgow not only made it through the University of Strathclyde, becoming the first member of his family to earn a university degree, he went on to get a Ph.D. in biology.

By 1990, Lithgow was working in the Biozentrum, the Basel, Switzerland laboratories of chemical giant Ciba-Geigy. He had moved up in the world, but now he was bored with his research. He became a frequent visitor to the Biozentrum's well-stocked library, where he looked for some new science avenue to explore. There, he read about the work of Tom Johnson, a University of Colorado biologist who discovered that some of his worms carried a mutation Johnson dubbed the age-1 gene. The worms with the mutation lived significantly longer than normal worms. Johnson had been trying to link genes with aging in worms since about 1982, but this was the first time anybody had detected a single gene mutation that conferred longer life in an animal.

"I said, 'Aging?' Nobody knows anything about aging," Lithgow recalled. When he typed the words *genes* and *aging* into a library database, the same few names kept popping up: Johnson, Sam Goldstein, fruit fly researcher Michael Rose, one or two others. Because aging was such an empty field, it was an attractive one. If a young, eager biologist were willing to defy the advice of just about every other scientist and pursue studies in aging, he might be able to make an impact quickly. So Lithgow began courting the handful of biologists who studied it.

On his first trip outside Europe, he flew to the United States to visit Goldstein's lab and, while he was there, received a notice that Johnson wanted to see him. "I went up to the counter at the airport and said, 'I want to go to a place called Boulder. It's in Colorado.'" After Lithgow found his way there, he spent several years with Johnson before returning to the United Kingdom to teach. He came back to the United States when he joined the Buck in 2001.

While in the United Kingdom, Lithgow was contacted by an old friend from Johnson's lab, Simon Melov. Melov, an avid sci-fi enthusiast with an affection for Lazarus Long, asked Lithgow to test a drug manufactured by a tiny biotech company. The drug was designed to be a potent free radical scavenger and the company hoped that it might help treat tissues damaged by radiation or chemicals.

When Lithgow used the drug on his worms, the worms lived 44 percent longer than other worms. The worms had not been genetically engineered. They were not calorically restricted. They simply took a drug and lived longer. Better yet, worms that had been deliberately damaged by potent free-radical-inducing chemicals recovered instead of dying. When the paper announcing the work was published in *Science*, it became the first time that the simple application of a drug increased an animal's life span. Both Lithgow and Melov, who was by then also working at the Buck, were inundated with requests from people hoping to use the drug on themselves.

This was a surprise to Lithgow, who was unaware there was such a thing as an "antiaging movement." His naiveté was partly the result of growing up in Scotland and partly the result of spending most of his adult life in a lab. But now he was sitting in a white palace on a hill in Marin County, California, where people cultivated an enduring faith in the perfectibility of human beings through hot tubs, or yoga, or alternative medicine, or modern technology and science, which made him something of an unlikely—and unwilling—star to longevity aspirants. Lithgow was cautious because he believed aging was part and parcel of complex life processes. If you interfered with aging, what else might you disturb? There was probably some sort of cost, a trade-off to be considered. There were problems with some long-lived worms. They were probably weaklings in the big boxing ring of evolution. Put them up against old-style worms, let them duke it out for food and sex, and the long-lived worms were harder to take care of in the lab. But he was slowly coming to grips with the idea that some people, hoping for salvation right now, were not going to be dissuaded.

While he tried to discourage false hope, the worms were a fact he could not ignore. And in the basement of the institute, there were mice—mammals—that appeared to be living longer, too. How could a life extensionist *not* be encouraged?

Lithgow sighed heavily, sucked air in between his clenched teeth. He was very conflicted. True, he said. The experiments in his own labs were now closely paralleling things the fringe had been saying for years. That the mice were responding to antiaging interventions, well, that was indeed something. Even if the drug used on the worms proved unworkable for people, the principle had been established. Aging could be dramatically slowed. The fringe just wasn't fringe anymore. Even as Lithgow admitted it, he was profoundly uncomfortable.

University of California San Francisco's Cynthia Kenyon, on the other hand, felt much more at ease with her antiaging work. Like Haseltine, she did not worry about fringe versus not fringe or her science reputation. That was already assured. And she was optimistic that life span extension did not incur a cost. "All I am saying is, it must not be the case there is some awful trade-off or we would not have long-lived animals." Of course there could be cost-free life extension. She had done it with her own hands and was positively jumpy over the prospect of doing it for people. She had been excited long before many other biologists dared even speak about it.

Kenyon was a quintessential worm person. A lover of animals of all sorts—she had once considered becoming a vet, and often took her dog to her favorite Italian restaurant in San Francisco—Kenyon lived and breathed worms. She first became interested in aging when she saw old worms. "I was working with these worms and after a month I looked at my plate and was stunned because you could see it. I had never thought of it. Oh my God, worms get old. I thought it was so sad, you know? Worms get old. It was very moving."

Kenyon, a tall strawberry blonde with an excitable manner, had gone from the University of Georgia directly into graduate school at MIT, where she worked with some of the best molecular biologists in the world. Like most others, she experimented with *E. coli* and phages, made one or two significant discoveries in the field of DNA damage and repair and the ways in which genes are turned on. But she wanted to work in whole animals so, she said, "I threw myself at Sydney Brenner's feet."

She was already on a science career fast track, but by moving from Cambridge, Massachusetts, to Cambridge, England, she assured herself of long-term stardom. She stayed in Brenner's lab for four years, again making big discoveries. After she moved to UCSF, Kenyon decided to use a worm mutant called daf-2 to study part of the worm's life cycle, a kind of hibernation phase called dauer. In the course of her study, she serendipitously discovered that the daf-2 mutants, when coupled with the activity of a second gene called daf-16, did not just live 40 percent longer as Tom Johnson's age-1 mutants had. Instead of dying by twenty days, these worms doubled their life span to forty days, twice as long as regular worms. She was captivated. "You look at these worms that live twice as long and it just takes away everything you always thought about aging. We always thought it was just the way it was. It was not negotiable, you know? But it is! And that's what's really cool."

"This life span extension, the largest yet reported in any organism, requires the activity of a second gene, daf-16," Kenyon wrote in *Nature*. "daf-2 and daf-16 provide entry points into understanding how life span can be extended." (It is this finding, and several since, that have led some in biology to whisper that Kenyon could one day win a Nobel Prize. She has already won the King Faisal Prize for medicine, collecting her award in Saudi Arabia along with the two other science winners that year, Craig Venter and biologist E. O. Wilson.)

"They should be dead. But they're not," Kenyon said of her worms. Big-time biologists generally did not talk like this. They herded their prized research into elaborate caveat fencing, caged it with rails of digression, until the bottom line became lost behind a corral of complexity. Kenyon reveled in the wonder. "It's really cool!" was a favorite expression and when she saw her worms, she thought she was onto the coolest thing ever.

"I wanted to get started quickly because, look, I'm getting older. Seriously." She had much more confidence than Lithgow that human beings really could intervene to slow or stop their own aging because worms were, in many ways, like people. "That's the big lesson I learned from developmental biology. All animals get old. All old animals look the same, they all get diseases of aging."

Over the years Kenyon and her lab workers progressively made worms live longer and longer. Eventually, they tapped into a genetic pathway, a series of domino-like events that follow one after the other, related to the system of insulin production and a gene called igf–1 (insulin growth factor), a gene that does for worms about what it does for people, who share a very similar gene. It was this research connecting insulin-related systems to aging that convinced Kenyon, a lover of Italian food, to stop eating pasta, bread, starches, all life extensionist no-nos.

She regarded her lab's igf-1 experiments as mindboggling. When she gave presentations about it, the excitement in her voice bubbled over and she flashed cartoony images onto the screen of a sad wrinkled worm supported by a cane staring with anticipation into a pool populated by worm brethren splashing happily in a spraying fountain. "Can *C. elegans* lead us to the fountain of youth?" the caption read.

Scientists who saw her talks always seemed to be waiting for her to justify the boldness. So she went on to explain how her lab used a new technique to interfere with the actions of genes to silence them selectively. Then they worked backward, tracing the effects of the silenced genes and what would normally trigger them to turn on or off. It turned out that environmental cues have a lot to do with the igf-1 pathway. When the team destroyed particular cilia, tiny, hair-like structures the worms use to smell food, they lived longer. Kenyon knew how crazy this sounded, but there was more. If you killed a gonad in a daf-2 worm to prevent the production of igf-related growth hormone, then interfered even more with daf-2, you could make a worm live six times longer than it was supposed to. At 120 days, half the worms would still be alive, like average human life expectancy rising from, say, 80 to 480 years. (By 2003, work by another team showed that impressive results were also obtained in mice, mammals that share most of their genes with people.)

The gonad killing, the cilia removing, the gene interfering, all tied in with germline stem cells and development through a circuitous route. "It's pretty cool. We're not getting four times longer, we're getting six times longer . . . the thing is, it's the germline. This is something Mike West would say, but it's true. The germline is totally immortal and that means life does not have to age."

Kenyon was not particularly worried about sounding too fantastic, partly because she believed this *was* fantastic, but she understood that she was bumping into one of the most basic of all human perceptions: people get old and then they die. Kenyon was arguing that aging could be uncoupled from death. She knew just how mind-bending the impact could be.

"It's interesting, you know? I am a very rigorous, critical scientist and yet the kind of things I look at are things you think should never happen. It should not be possible. And yet there it is, proving you can come face to face with miracles."

And where there are miracles, there must be a company.

Kenyon first thought of an antiaging company in the early 1990s, the moment she first looked down at her plate of lively, squiggling mutant worms. "I wanted to start a company right then. I mean, you look at these worms, the mutants, and you say, 'If you were a worm, you'd rather be that one.' It's really visceral . . . I wanted to be those worms."

The company never got off the ground, but in 1997, an MIT biologist named Leonard Guarente, who was doing for yeast about what Kenyon was doing for worms, felt his technology might be ready for the corporate world. At the time, he was consulting for a company called DeCode Genetics, a research firm that studied the genes contained within the population of Iceland as a way to track down disease-causing, or health-giving, genetic types. Guarente approached one of the VC founders of DeCode, Cindy Bayley.

Guarente could not have picked a more enthusiastic target. Bayley was an admirer of Wally Steinberg. She had read the 1992 *New York Times* story in which Steinberg proclaimed that he wanted to start a company devoted to aging and was intrigued. When Steinberg died before he could finance one, Bayley looked for a way to do it herself. By the time Guarente approached her, she was already sold.

After her conversations with Guarente, Bayley thought any future company should have a *C. elegans* component and that Kenyon ought to join the team as a scientific co-founder. Kenyon readily agreed. In 1999, they incorporated, calling the new company Elixir Pharmaceuticals, a shameless, in-your-face moniker that dared the biology establishment to make fun. Not many did. The establishment was beginning to think that aging might be crackable after all.

By 2000, Elixir had the funding it needed and a founding inspiration, Shakespeare's poignant sonnet number 60, a poem with a message-in-the-bottle resignation that expressed a hope in the survival of art, if not man.

> *Like as waves make towards the pebbled shore,*
> *So do our minutes hasten to their end;*
> *Each changing place with that which goes before,*
> *In sequent toil all forwards do contend.*
> *Nativity, once in the main of light,*
> *Crawls to maturity, wherewith being crowned,*
> *Crooked eclipses 'gainst his glory fight,*
> *And Time that gave doth now his gift confound.*
> *Time doth transfix the flourish set on youth*
> *And delves the parallels in beauty's brow,*
> *Feeds on rarities of nature's truth,*
> *And nothing stands but for his scythe to mow:*
> > *And yet to times in hope my verse shall stand,*
> > *Praising thy worth, despite his cruel hand.*

Any company that says it wants to retard the aging process with drugs faces a problem. If you want to say you make a drug that slows aging, the test to prove effectiveness would last for decades. So Elixir developed a strategy for attacking the symptoms of aging like diabetes, cancer, arthritis, and heart disease. It was going to try to create drugs that would prevent, slow, or ameliorate these symptoms of aging the way statin drugs were used to lower cholesterol. High cholesterol was not a disease. Statins had been approved because they lowered cholesterol and, by implication, prevented cardiovascular disease. Elixir hoped its drugs would be approved by a similar route and that the side effect would someday be proven to be the slowing, or even the stopping, of the entire aging process.

Ed Cannon, Elixir's CEO, an old Harvard acquaintance of Haseltine's and a one-time business collaborator of Human Genome Sciences, admitted that though the strategy sounded reasonable enough, some of his colleagues were pretty sure he had stepped off the deep end when he joined the company. But like everyone else who advocated life span extension, he was an optimist. While the bio-Luddites pointed to the comforting way death

cleansed the world of evil by cutting down madmen and cruel dicta-
tors, and used this as an argument for banning technologies that led
to longer maximum life spans, optimists like Cannon concentrated
on the flip-side. "The quality of life will be improved, and for us,
that is clearly a less controversial message than talking about how
much longer you will live . . . We look at this simply. We have to age
because the only way to reproduce is to mature to reproductive age,
but once that has happened, it would be nice to modulate aging as a
way to avoid disease or impairment to our quality of life. I think that
is a good thing. If we live longer, so what?" Would it not be nice,
Cannon argued, to have some brilliant people around? "If Albert
Einstein were alive, look at the problems he might be able to help
us solve. Or any of the great humanitarians? Would it not be great if
Gandhi were floating around India, Palestine, Afghanistan? We lose
those assets prematurely."

This was a perfect echo of a familiar transhumanist argument for
life extension. The transhumanists weren't just afraid of death, they
were pissed off by it because death was a pickpocket on a grand
scale. "Death is an outrage!" nanotheorist Robert Freitas liked to say,
the greatest catastrophe to ever hit humankind, a $100 trillion loss
every year, an ongoing economic disaster. We ought to be living at
least 1,100 years given our science prowess, and investing in making
sure we do would be worth it because every person had a value of $3
million to $7 million counting everyone's productivity, accumulated
wisdom, talents. Imagine how much the great thinkers would be
worth. But each year in the United States alone, death robbed us of
the equivalent of three complete Libraries of Congress.

This was not the kind of economic impact those entering the
aging space thought about, necessarily. They looked at revenues. If
somebody did come up with a drug that stopped aging, or even slowed
it down a little, that drug would instantly become the biggest block-
buster of all time, in a market that would include every human being.

Lithgow's experience with all the calls and e-mails from hopeful
test subjects following the drug experiment proved that forecast. And
when Kenyon talked about her research in public, she was often
approached by audience members asking if they could, please, be
guinea pigs in whatever life extension study Kenyon was about to do.
Once, a man demanded he get access to any new drugs as soon as

they hit the development stage. "It was kinda scary," Kenyon recalled. "It was like I had gold and nobody else did."

But she would rather people be enthusiastic hopefuls than doomsday prophets. She had no patience for those, like Fukuyama and the bio-Luddites, who objected that life extension is somehow immoral. What was the point of biological research? What was the point of all those summer meetings in the shabby cabins at Cold Spring Harbor after the war, of figuring out the structure of the double helix, of the countless experiments with *E. coli* and phages, of the Human Genome Project? What was the point of the billions of tax dollars spent every year by the NIH? Why not just let cancer run free? Why wage war on it? "The whole reason that we do what we do with the taxpayers' money, is they want something from it. That's the whole point. It makes me a little angry to think we should be in some ivory tower learning how life works supported by the taxpayers. What's that about? They want something! That's why we do it." If people see her work as a form of salvation, "I think that's good! That's why they pay us. That's the whole idea. Otherwise why would Joe Schmoe allow his tax dollars to go for medical research?"

IF PEARSON OR SHAW used the word *immortality* they were written off as cranks. When the Extropians said that human beings could have a "transubstantiated future," few outside the circles of the alternative bio-techno-utopians listened. But Haseltine was the CEO of a big-time biotech. He had homes in Georgetown and New York, the wife, the Harvard professorship, the membership in the ultimate insider club, the Brookings Institution, and along with Kissinger and much of the rest of the world's power elite, a seat on the Trilateral Commission. Inoculated from the charge of being kooks by prestigious degrees, scientific publications, and money from venture capitalists, scientists like Haseltine, Kenyon, Dominko, and West had turned the natural weirdness of biology into biotechnology, and no matter how crazy it all sounded, the next step to a future paradise was almost here. Christopher Reeve would walk and Michael J. Fox would be cured and there were stem cells, and genetic engineering and smarter kids—and immortality!

Of course it would come true, no matter what the bio-Luddites said. Americans have always had an insatiable attraction to promises

of longer youth and better health. The age of the gene and biotechnology did not change that, it just expanded the vocabulary. "When
they're not busy cloning, today's DNA scientists are engineering
high-tech skin creams," declared a breathless article in *W* magazine
on a company claiming to custom design "clarifiers, moisturizers,
and anti-aging serums" according to individuals' genetic sequence.
"If you're not genetically programmed to make enough collagen,
Lab21 will make it for you. The treatment will sell for 250 dollars and
be shipped in less than a week."

Pitches like that succeeded because making one's life better is a
natural human desire. Far from frightening some people, the
prospect made them say, "Hey, that sounds pretty good. I'd like a bit
of immortality, a better memory, a younger look."

Donna, for example, was already trying. As early as 1997, when
she stood by the calf thymus extract booth at a dietary supplement
trade show, she was an early adopter. She was in her middle thirties
and on a quest to find the latest, best, most effective transforma-
tional techniques and products science had to offer to extend her
youth, improve her body, make her what she hoped to become.

She was tall—about six-feet in her heels—fit, busty, and dressed
for attention in tights and a low-cut blouse. Her lips were fluffy as
down feathers, her blonde hair cascaded down her back like a water-
fall. She wore sunglasses though she was indoors. She was also very
popular. Wherever she stopped on the show floor, a huddle of men
alighted, surrounding her with awkward smiles, proffered business
cards. When she walked, the clack of her stilettos churned up a
foamy male wake.

It hadn't always been like this for Donna. She was sickly in her
twenties, suffering from an ongoing battle with endometriosis and a
host of other, more minor, complaints. She felt miserable, then. Her
body and her self-confidence were broken. She hated feeling that
way and vowed she would do all she could to make herself invulner-
able. So she decided to model herself on the idealized art of Helmut
Newton and his photographs of extremely long-legged, high-heeled,
dominant women of Amazonian proportions. Those women were
confident, sexually powerful, iconic. Donna was already well on her
way to achieving the Helmut Newton ideal but as successful as she
was, it wasn't enough.

Even Newton's women aged just as Donna did over the next few years. They were susceptible to disease. They were human, and being like them, as good as it felt—and Donna thought it felt wonderful—was just the last completely human stepping stone before she crossed over. Someday, she hoped, she might be truly unassailable by sickness, age, death. So she set her aim on another goal, the transcendence of Hajime Sorayama's "gynoids."

Sorayama, a Japanese painter, was famous for creating hyperrealistic images of genetically manipulated gynoids and "sexy robots," women who were a cross between the female robot of *Metropolis* and a latex fetishist's ultimate fantasy. They were tall, lean, often sheathed in metallic coatings with the exception, usually, of their breasts, which ballooned from powerful chests. Orifices were modified, metallicized, glorified. Gynoids were bullet-proof, disease-proof, age-proof figures of devotion, lust, power. Donna had a collection of Sorayama's prints on her wall, his books on a coffeetable. She understood she was not about to obtain gynoidom, but she used the images to inspire her vision of herself, provide motivation for her quest. And then, if the technology advanced, she vowed to do whatever it took to get as close to gynoid as she could.

"I've traveled all around the country learning about stuff I want to do," she said. "The treatments, the places, the doctors, the therapists." Thanks to her marriage to a wealthy man, she was able to make this her full-time job, spending about $1,000 per week. She was a regular at spas, and not just any spas, but the Saturnia at Doral in Miami, the Golden Door in San Diego, Canyon Ranch in Arizona. Technicians came to her home, too. "You could have a pretty good party with all the people who work on me," she said, laughing.

"How many people do you suppose?"

She paused a long time, counting.

"In all, I think eighteen."

Botox for her face, Restylane for her lips, a massage therapist, a cosmetician, a hair expert, a nutritionist who gave her IV drips of vitamins and minerals, a special antioxidant infusion that cost $250 per treatment, doctors to prescribe hormones. She knew where to find the most advanced of everything, whether it was FDA-approved or not. She was auditioning plastic surgeons. She wanted them to get a good look at her body now, while it was still remarkable, so they'd know what to do later. She worked out every day, slept at least eight hours, and changed her hair color on a whim. "I get bored looking at myself. I've had red hair, bronze hair, blonde hair, honey-colored hair." Her body was truly her art and her temple. The one thing she had not been able to do yet was genetic modification. She asked where she might go. "You want me to fly there tonight? I'd do it. I am definitely an experimenter."

As the years passed, a divorce from her husband clipped Donna's wings only slightly. It had nothing to do, she said, with her quest. The settlement was generous, and while she had to cut out the IV drips, and she stopped going to spas altogether, dropping her esti-mated weekly expenses down to about $600, the rest of her program was as it was. She was successfully holding age at bay, she reported, and dating men ten years younger. Still no genetic modification, but she hoped the day was coming soon. If she could, she'd turn herself into a gynoid tomorrow. She knew she couldn't. She knew she might not ever be able to. But the idea of her transformation continued to animate her because it was a thrilling beacon of the future.

The idea of human transformation was not restricted to people like Donna, or the rare scientist like Haseltine. Mike West was a long-time transhumanist buddy. Other well-known scientists started to socialize with Max and Natasha, too. In 2001, for example, the couple held a party in celebration of Primo 3M+, Natasha's web-based conceptual rendering of the body of the future, one equipped with a wide variety of mechanical, digital, and biotechnological improvements. The ideal future body, as it happened, looked a lot like a silvery, airbrushed Natasha, but it included a waste recycling system, telescopic vision, turbo-charged brain implants, Technicolor skin, vastly improved sensory ability, and about a dozen more upgrades.

The debut mostly attracted friends, including Extropians, fellow travelers, and a few pals who had nothing to do with transhuman-ism, like low-level L.A. film industry types. A small coterie of "CR" dieters showed up, people following calorically restricted diets in an effort to slow their own aging. They sat on a couch nibbling the edges of tortilla chips.

Then in came the scientists. Michael Rose, the University of California Irvine fruit fly scientist Lithgow discovered when he looked up "aging," was among the early arrivals. Rose achieved biology fame by breeding flies that lived extraordinarily long lives, far longer than fruit flies had ever lived before, one of the first experiments to demonstrate that aging had a genetic basis. A few minutes later, Roy Walford, the scientist who popularized caloric restriction as a life extender, entered on shaky legs supported by a friend. (Some months later he would reveal that he suffered from Lou Gehrig's disease.) Both men, along with transhumanist icons like Ralph Merkle, Moravec, Minsky, and Freitas, had helped Vita-More design her future body. At one point in the evening, Rose, an admirer of Haldane's, talked about one of Haldane's last great presentations, on the future of life in the next 10,000 years. He took a sip of wine and thought about the speed of biotech. Ten thousand years may have been pessimistic. "Who knows," he said, "maybe Natasha will get her future body after all."

9   Pop! Goes the Rapture

JOHN DRAPER STARED DOWN into his confocal microscope. "It's too bad you weren't here," he said. He looked up with a fish tale, shoulda-been-here-yesterday face. "We had heart cells beating in the dish."

Peter Andrews smiled, and confirmed that yes, there had been some human heart cells beating in a dish, a pumping cumulus cloud, ka-choong, ka-choong, ka-choong. But don't worry, Andrews said soothingly, it was not some biotech breakthrough. Embryonic stem cells often seem to want to become heart cells, called cardiomyocytes, all on their own and neither Andrews, nor Draper, his graduate student, nor anybody else for that matter, knows exactly why. They just do it. It's kind of amazing when they do it, which is why stem cell scientists bring videos of the beating cells to science meetings. The rhythmic pumping is a dramatic illustration. An illustration of what, however, is open to debate. A few biologists leave the impression that the beating cells illustrate their own great skill at transmogrifying ES cells from lazy circles of protoplasm into vibrant tom-toms. Others say that they had no intention of making heart cells in a dish. The damn things just appeared. Some say they didn't want cardiomyocytes at all, that it was a mistake, but ES cells are so itchy to become something, anything, that they are difficult to control.

Andrews falls into this last category. He's been around cells like these for thirty years or so, longer than most other scientists who work in this area of biology, and so he's seen the rise and fall and rise again of expectations for miracles. Now the expectations were at an all-time high but Andrews retained a cool detachment when he talked about them. He had no intention of turning ES cells into heart cells. He knows, as does every other developmental biologist, that beating is one of the first things embryo cells do and they'll do it whether you try to make them or not.

Draper returned to his microscope, scanning for clumps of ES cells. He found one, and then Andrews leaned over the scope to confirm the sighting. I sat down for a peek. Sure enough, there they were. Amazing how much seemed to be riding on the backs of these not-so-impressive circles. Depending on whom you asked, they were either the salvation of man, or the paving stones on the road to Perdition.

I looked up from the microscope at Andrews. "That's it," he said, laughing. "That's all there is to see."

He laughed because I had made a pilgrimage all the way to Sheffield, the old industrial heart of Yorkshire, a Victorian town of red brick buildings etched along winding roads that lead up a hillside from a narrow valley. The lengths to which some people would go to look through a microscope and see some cells, I imagined him thinking. He was a scientist of long experience who had suffered much frustration in his career, so when he looked in the dish, he saw cells. Interesting cells. Even fascinating. But just cells. He was not oblivious to the possibilities, the visionary prospects, the fears. He even thought about a transhuman future, sometimes, when he went off to the pub to share a pint with friends or in the dead of night when his imagination was free to roam. "One does think about it," he said. "One does think about all sorts of weird and wonderful things." But like many scientists, especially those who worked in the emotionally charged field of embryonic stem cells, he tried to resist making big statements, preferring instead to concentrate on getting the damn things to just live in a dish.

Yet the very existence of the cells in a dish in Sheffield already testified to their power as symbolism. They had come from James Thomson, the University of Wisconsin scientist funded by Mike West's Geron as part of West's own attempt to defeat death. As such the little clouds of cells had become infused with the power of faith, the most crystalline manifestation ever of a new religion, the biotech religion, that had been building in the mind of man for hundreds of years. Wells gave it expression. Haldane and Schrodinger proclaimed it possible. The early molecular biologists pulled the tools to achieve it out of nature's box. Biotechnology and venture capitalists used the tools and stoked the faith of the futurist subculture with revelations of a coming paradise. And now the embodiment of the new religion rested right here in this dish.

What made this new religion a religion rather than a technology was that Peter Andrews, and every other developmental biologist, and every geneticist, physicist, and capitalist, too, were stumped on how to make the bioutopian future a reality. The rapture was turning out to be much more difficult to call forth than Wallace Steinberg and many others said it would be. So the belief in the coming age of bioutopia was still that: a belief.

Peter Andrews knew this better than most people. By the time he was taking graduate degrees, the British specialty of developmental biology was being eclipsed by the glamorous gene-centric molecular biology which was largely an American affair. This eclipse took on an uncomfortable literalness. In 1974, during the bad old days of British economic decline, Andrews was sometimes forced to write his Ph.D. thesis by candlelight when Oxford, like much of Britain, suffered periodic blackouts caused by one of Britain's coal miner strikes. Now the government of the United Kingdom was attempting to step into the gap left by American hand-wringing over stem cells to build what it hoped would be an economic windfall based on the technologies of ES cells and nuclear transfer (cloning).

But Andrews, a tall, genial fellow with salt-and-pepper hair, who is much less high-strung than some of his colleagues in the United States, prefers to skip all such wide-lens views of ES cells. He sees them as the logical closing of a circle he helped draw since the start of his career, a period that saw his field become a biology backwater, then, two decades later, marqueed as the hottest thing in biotech since gene splicing. The fact that the cells in his dishes came, through a roundabout route, from Mike West, only made the trip that much stranger.

By the time Andrews saw the beating heart cells in his dish, almost three years had passed since Thomson had announced his isolation of human ES cells. It had been four years since Dolly's debut. But Andrews seemed somewhat removed from the worldwide fuss. As he walked out of his lab building, then down the street to a small pub frequented by university professors, he said he had a little interest in a small biotech company that was based on his ground-breaking studies of embryonic carcinoma—teratoma—cells, the cells from Bizarro World that made inside out and backward embryos. He didn't seem especially excited about cashing in, though, and he was surprisingly diffident about looking forward. He kept referring to his career and the events of the 1970s, even to the work of Gurdon in the 1960s.

Over lunch and beers, Andrews talked history. To an outsider, it would appear that much had changed in his field, but Andrews viewed the advent of human ES cells and cloning differently than most of the media and far differently than legislators. To him, ES cells were most valuable as a way to put research back on the track he had hoped to pursue in the first place. To him, most everything that has followed Dolly and the isolation of human stem cells amounted to so much noise. He pushed his beer away and leaned back in his chair. "When Jamie [Thomson] finally succeeded in get-ting human ES cells, suddenly it became interesting again to use ES cells to study the process of cell differentiation. This takes us back to the 1970s."

We talked about the twists and turns of science, the characters, the blind alleys, the inevitable false starts. This is the way science worked. You just didn't say you were going to create human immor-tality, an ultrahuman future, and then make it happen. You had to make a lot of mistakes—mostly mistakes—and then, finally, you fig-ured something out. That's the way it had always been, and while new technologies may have sped things up a little, that was the way it would always be. Andrews sounded almost relieved to say that the whole idea of using embryonic stem cells as a therapy was overblown. "I caution, there is an awful lot of hype," he said. "Using stem cells—in whatever form—routinely is quite a long way away. I say, I seriously doubt there will be any treatments [using stem cells] for anything for fifty years." He took a deep breath and rattled off a list of practical hurdles that had to be overcome. Getting the damn things to simply live was tricky. Andrews had been working with the ES cells James Thomson had sent him for eighteen months, and it had taken him that long just to keep them from croaking or spontaneously differentiating into heart cells or some other damn thing. When Andrews tried, he had trouble developing his own ES cell lines. He used spare embryos from IVF clinics, struggled to tease out the ES cells and make them grow, but though he had been at it for over a year, as of early 2003, all his efforts had met with failure.

But what of the dramatic videos, the published papers in journals with photographs of stem cells becoming nerves and skin and organ cells? "The gist of what is published and what is said at meetings is 'Look at this wonderful cell line, and look at this lump of beating heart muscle or look at these nerve cells that come out.' That's nice, but on the whole it has not told you very much about how you go from one to the other."

It wasn't that Andrews didn't think human embryonic stem cells were important. He did, and he thought the obstacles placed in the path of research on them was harmful to science and based on religious objections that made no sense at all. It's just that so much had been made of them, as if they really were like a drug you could take to cure what ailed you. But Andrews said there was very little evidence that ES cells would provide any of the miraculous advances science had advertised, at least not now. "We are not about to make Christopher Reeve walk," he sighed. Andrews wasn't alone. Other developmental biologists began to air their own misgivings, even predicting a backlash when the public realized that there had been far too much promise and far too little delivery.

This is exactly what happened and the backlash was not only over the "promise of ES cells," but over the entire biotechnological prospect, as those who had been primed for the coming rapture began to realize it wasn't coming any time soon.

HASELTINE'S PARTNERSHIP WITH Venter's TIGR fell apart in June 1997, as many had predicted it would. "In a not-too-surprising announcement . . ." began the notice of the split in *The Scientist*. The HGS press release thinly glossed over just how corrosive the relationship had become. "Both parties agree that after almost five years of collaborations, separation of future research is in their best interests."

In fact, the two miserable spouses in the shotgun marriage arranged by Wallace Steinberg were both looking for a way out.

Haseltine had resented the deal from the start. He had his academic acclaim, his Harvard tenure, and he did not see why he had to fork over millions of dollars every year to TIGR out of the HGS budget to salve Venter's insecurity about his place in the science firmament. This was one of the reasons why Diversa (originally Industrial Genome Sciences) was started. "The motivation for Bill," CEO Short said, "was that he was having to pay $8 million a year to TIGR as part of this deal that Healthcare Ventures set up. He wanted to offload that and have us pay $2 million of it" to TIGR so that TIGR could sequence microbes for Diversa. The aborted Plant Genome Sciences, which was to have been part of Steinberg's plan to become the Standard Oil of genomes, would have paid another $2 million.

Venter wanted rid of Haseltine as much as Haseltine wanted rid of Venter. He was not any happier with the deal Healthcare had set up, a deal that, as much as it mortgaged HGS's money, also restricted TIGR's ability to become the genetics skunk works Venter imagined it would be. "Craig Venter, head of TIGR, says HGS's threat to get a court injunction to prevent TIGR from going to press with the sequence of the *H. influenzae* genome demonstrates how bad the situation had become," reported *Nature Biotechnology*. "HGS's CEO Bill Haseltine says that none of this ever happened."

"We were feeling more and more constrained by all our rights going to this one biotech company," Venter told *The Scientist*, subsequently telling the same magazine that HGS was "holding data hostage" and "keeping it secret." "Scientists can't have access to it," he complained.

Short had another spin. "Craig wanted his independence."

Then, one year after the breakup, Venter did what he said he did not want to do. He jumped into private enterprise, with both feet this time, using the patronage of laboratory equipment maker Perkin-Elmer to start a new company, Celera, which declared that it would sequence the human genome faster and more cheaply than the multinational public effort, and that it would sell this information to pharmaceutical companies looking for drug targets.

None of this inside back-and-forthing mattered much to HGS or to Haseltine because by the time the biocentury swell rolled in, Haseltine and Human Genome Sciences caught the wave perfectly.

Outside the HGS walls, an insatiable hunger for a piece of the bio-future was driven by the near completion of the Human Genome Project, by magazine covers announcing the dawn of personalized medicine, an era in which we'd all have drugs designed for our own special genomes. Just about any scientist worth his or her salt was on a corporate advisory board, had started a company, was talking to a venture capitalist. Investors fell in love with biotech as they had not since the days following the Genentech IPO. The biotechnology index rocketed more than 300 percent in 2000, raising about $30 billion. As one of the biggest and the first of the genomics-based companies, Human Genome Sciences found itself surfing in the curl. HGS had deals with big pharmaceutical companies, was entering real drugs in clinical trials.

Haseltine was fearless in making the most of the opportunity. Throughout 1998 and half of 1999, Human Genome Sciences' stock price traded in a narrow price range for a biotech (though it would have seemed much wider for a blue chip), moving side to side, mostly. But by 2000, at the height of the flurry over the human genome, at a time when talk of changing human biology forever was on nightly news broadcasts and the subject of political debate, Human Genome Sciences began a steady climb. On January 31, 2000, HGSI split two for one, but the split barely made a ripple in the stock's smooth ascent upward. One month later, on March 1, HGSI closed just above $112. Haseltine justified the price by promising drugs soon. The first ones would hit the market in 2002, or 2003 at the latest, he said, and they would "be the first of many." Genomics was going to pay off in a big way, and though there were other genomics companies, Human Genome Sciences would prove to be the dominant force, because most of the new drugs found through the study of genes would come from his company.

On September 28, HGSI closed at $90 per share and a week later the company again went back to the market for more money, splitting in another two-for-one deal on October 6. By October 18, the shares had risen again to just over $100. Thanks to his repeated forays into the market, by the end of 2000, Haseltine was sitting atop a cash wad of about $1.6 billion and had secured a reputation as the leader of the company that would ride point for the rest of the biotech pack in creating a new set of human drugs and information that would change medicine forever.

Haseltine was so confident that HGS would be producing a stream of new drugs that he used some of the corporate bankroll to build a new campus, complete with all the manufacturing he could want. Repifermin and the other drugs being tested were made in a small pilot facility in the basement of a building HGS leased in Rockville, Maryland, but the new HGS headquarters, rising in a muddy field several miles away and shaped like a giant Y to mimic the shape of human antibodies, was scheduled to open up in late 2003.

Some in biotech whispered that Haseltine was crazy for investing so much in a new corporate campus when he didn't even have a drug ready to market. It was a testament, they said, to his ego. They sniggered behind his back in anticipation of a fall. But doubts, criticism, quibbling only gave Haseltine fuel. The more criticism he took, the more driven he became. He always performed his best, and enjoyed life the most, when proving the conventional wisdom wrong and he knew that biotech had been built by audaciousness, something he had plenty of. From the time he was a boy back in China Lake, he had been trying to prove to the world that he had the biggest stones and the biggest brain around, and his claims were just more in a long line of truth or dare statements.

But Haseltine and all of biotech got a shock when both Celera's and the international consortium's drafts of the human genome were published in February 2001. Both said human beings had somewhere around 30,000 genes. For years, geneticists had estimated that people had anywhere from 80,000 to 100,000 genes. When HGS marketed its data to big pharmaceutical companies looking for drug targets, it said it could deliver something like 90,000 genes. With the publication of the genome drafts, it looked like pharma had been buying fiction.

Haseltine jumped on the results, re-creating the old head butting battles he so enjoyed at Harvard. Once again, he insisted that the recognized experts, Venter and Francis Collins, the men who had stood by the side of President Clinton as he hailed them in the same terms presidents used for the first astronauts, had gotten it wrong.

He had never liked the idea of sequencing the whole genome. He thought it was a waste of time and he irritated the sequencers by saying so at every opportunity. Now they enjoyed telling the world that Haseltine's company was worth a third less than he had been saying, and Haseltine enjoyed telling the world that the famous gene sequencers had botched the count. Scientists and biology watchers chose sides depending on what they thought of Haseltine and the genome projects. Haseltine was either a whiney loser, or a brave soothsayer who was unafraid to go public with an opinion others dared to air only among themselves in laboratory snack rooms. When Collins was challenged with Haseltine's argument that the count was way off, he snapped, "He's the only one who thinks so!"

Well, not quite as it turned out. Over succeeding months, research teams around the world came up with new numbers and almost all of them were higher than the 30,000 or so the two genome projects counted. None were as high as Haseltine's, but he was partially vindicated.

Just how many genes people have doesn't really matter, at least not much. "Some people just like to argue," Craig Rosen, Human Genome Sciences' vice-president and head scientist said. "In fact, none of us know." Rosen was sitting in a conference room next to Haseltine's office. He cocked his head toward Haseltine's wall and lowered his voice. "He does not know. The [Wall Street] analysts do not understand. We say there are a lot of genes. Whatever we have is physically there. I really believe between 80,000 and 100,000."

No matter who was right, biotech investors were alarmed. Rosen said that nobody knew how many genes there were and while it may not matter so much what the number is, the fact that nobody knew made investors anxious because the scientists were supposed to be the smart ones. They were supposed to know. What else didn't they know? After one hundred years of unraveling the onion of biology, it only seemed to get more complicated. "One gene, made one protein, made one drug or one antibody target. It was very simple." That was Haseltine's mantra for Human Genome Sciences and the "Central Dogma" of molecular biology. One gene. One protein. A gene in the nucleus of a cell was "read" and transcribed by a molecule called messenger RNA. The messenger RNA carried the instructions to the ribosomes. The ribosomes made the proteins. The proteins made everything. They told your cells when to die and when to multiply. They made you grow, they healed your wounds, they kept your body humming along. So when a protein went missing, as in diabetes, it was vital to replace it. And if a little protein could, say, make your body heal itself, then more would make it heal faster and better, no? Or maybe something's gone haywire and a protein isn't being made correctly; it's tangled up in some weird contortion making cells do crazy, wild things like make more and more of themselves, their gas pedals stuck to the floor until they overwhelm everything in their path. That was cancer. If you could make the right antibody, you could shut cancer down. No wonder Human Genome Sciences could dip its ladle into the market and haul out money. It had the best story in biotech. But now it was clear that a lot of smart people like Haseltine who said some very definite things about "The Book of Life," did not know as much as they thought they did.

In fact, if there really were only 30,000 genes, then each gene must be capable of making a number of different proteins through a process called alternative splicing. And all those proteins were part of a ridiculously complex web in which each strand depended upon the presence, the strength, the amount of other strands. Even if Haseltine was right, and humans had 80,000 or 90,000 genes and the gene-to-protein ratio was closer to one-to-one as stated in the Central Dogma, the action of one protein could influence the actions of any number of others.

Biotech introduced a whole series of new technologies to address the complexity, each one hailed as the next great thing, a trendiness that moved faster than Deeda Blair's fashions. First genomics was the grand frontier. Then new companies were formed and they said no, the frontier was "proteomics," the search for relevant proteins, and then it became "metabolomics," until, finally, one publishing company threw up its hands and began publishing a journal called, simply, *Omics*.

By late 2001, biotech investors were waking up with a hangover and finding themselves in bed with a lover who didn't look so hot after all. People had bought the story Wallace Steinberg and other venture capitalists told, just like they had done when Al Scheid was creating a generic California Biotech, but now they wanted the story to change from "promise" and "potential" to revenue. They wanted to see some product, something, anything, besides the abracadabra of "genomics." Finding new genes was great. Way to go. Hurry up and do something with them, for chrissakes. At the February 2002 annual meeting of the American Association for the Advancement of Science in Boston, Jerrold Fuchs, a retired investment banker-turned-private investor, listened to a science talk about genomics. At the end of the talk Fuchs sighed, "I'm beginning to think this stuff is never going to work." The first drug Human Genome Sciences sent to clinical trials failed. The second, Repifermin, the drug that miraculously closed the wound in the video Haseltine showed at his meeting, was in serious trouble. There would be no HGS drugs on the market in 2003 as Haseltine had said there would be.

Haseltine can be funny, charming, and, when he chooses to be, gracious. Friends and foes past and present agree that he makes a great drinking buddy. But he has always been a driven man with unyielding ambition. Those who dislike him sharpen their knives on this trait as if Haseltine were the only freakishly competitive biologist with a soaring ego. His own sister, Florence, admires his brain, brags about his photographic memory, calls him a genius. In 1977 she said she thought he'd win a Nobel. But she also said he had a "difficult" personality. Veterans of the bloody academic arenas of Harvard and MIT have long memories and Haseltine made his share of enemies during his years there. Some who knew him back then kept their fingers crossed when he first took a three-year leave of absence to run HGS, counting the days until the three years ran out, he lost his tenure, and he reached the point of no return. Ten years later, the mention of his name brought out the long knives. It wasn't just that he was arrogant, self-aggrandizing, and manipulative. It was his way of dodging karma. He was all those things, but he still became rich, still became famous, still kept getting—and taking— credit as a special kind of oracle with his high-powered connections, his glamorous wife, his globe-trotting. A former Harvard colleague who watched his career with a perverse fascination said that Haseltine's trajectory inspired wonder. "One of the questions I ask myself about him is, When is it ever going to be enough? When will he satisfy himself that he does not need any more?"

Finally, some in the world of biology and biotech thought, it was their turn to bathe in schadenfreude. Even his old friend Ed Goodman, the man who started him in the biotech business by teaching him what an investment banker was, and Haseltine's old mentor,

James Watson, who lived near Goodman in Oyster Bay, New York, sometimes got together and sniggered over Haseltine gossip. "Bill is respected for his intellect," Goodman said, "but he is not loved in the community."

Biotech watchers, some of whom barely knew Haseltine, piled on. "Genomics doesn't work," one said with obvious pleasure. "Bill Haseltine's got nothing."

After HGS bought out a little company that owned a technology to increase the length of time protein drugs were active in the body, and began an aggressive campaign to link that technology to existing drugs like growth hormone, the critics pointed to the move as proof that the genomics revolution had floundered. Genomics was supposed to fill the pipeline, rescue big pharma, and now the leading genomics company was trying to make money piggybacking on other companies' drugs, trying to create improved versions to make a quick buck. "The bet was that they could get drugs to market faster than the industry benchmarks, and I think they lost the bet," a biotech consultant told an industry newsletter.

Human Genome Sciences was hardly alone on the slippery slide. Thousands of biotech companies had been started on the salesmanship of men like Wallace Steinberg and hardly any, maybe a handful, made a profit. The malaise was so severe that when Human Genome Sciences held its annual investors meeting in a ballroom of the Waldorf Astoria in New York City on April 30, 2002, the share price had been beaten down to less than $16.

The Waldorf ballroom was crowded that day, but when Charles Duncan rose to ask the first question of Haseltine, the room was silent. There was obvious skepticism in his voice. "When do you think the use of genomics information will improve drug development and not only discovery, and how will we see that in HGS' use of that technology?" Duncan, one of about 200 people at the luncheon, was a Wall Street analyst who happened to work for Dresdener Kleinwort, but the name of his employer could have been Merrill Lynch or Lehman Brothers or J. P. Morgan or he could have been a big individual shareholder, a guy like Jerrold Fuchs or just somebody who had gotten swept up in the religion of biotech, because all these people wanted to know the same thing. Not just "when would genomics work" but "when would biotech work"?

The industry laid off thousands of workers. Stock prices plunged. Headlines like "Biotech Sector Bleeds Money," "Companies that Seek Cures Now Fight for Life," and "Biotech Firms Face Funding Crisis," ran in major newspapers and business magazines. Venter left, or was booted out, of Celera Genomics, which announced that, like Human Genome Sciences, it now wanted to find and make drugs, a tardy acknowledgment that it was buying into the strategy Haseltine had set out for HGS ten years earlier. Gone was the hoopla of the human genome and Celera's role in it. "Surely this company, founded by genome-sequencing megastar Craig Venter (who's no longer with the firm), could someday discover a cure for cancer, but we think there's just as much potential to destroy shareholder value over the next decade," reported the investment ranking company, Morningstar. "I think I'd rather earn a guaranteed one percent on my cash in a money market account."

Thirty years after Theodore Friedmann proposed it in *Science*, sixteen years after Deeda Blair and Wally Steinberg plucked French Anderson's technology out of the NIH to set up Genetic Therapy, gene therapy almost never worked. In 2001, at the annual meeting of the American Society for Gene Therapy in Seattle, Inder Verma, Haseltine's old lab mate, and a French doctor exulted in the announcement that children in France who suffered from severe combined immune deficiency syndrome (SCIDs), had apparently been cured. At last, there was an unqualified success. A year later one of the children turned up with a rare form of leukemia, followed several months later by another one. The virus used to carry the gene into the genome of the cells had apparently lodged in a bad spot, a place that could trigger cancer. Gene therapy trials were stopped until science could figure out what had gone wrong.

The world of ES cells and cloning was roiled, too. Science was slowly deciding that the future was not going to be about cloning, it was not going to be about "therapeutic cloning," and it was not going to be about embryos or even ES cells. There was still a lot of science to be learned from all these and biologists were passionate about their importance as subjects of study. But in the political battle over ES cells and cloning technology, biologists and biotech had overreached, just as the Human Genome Project and genomics had overreached with all its "Holy Grail" and "Book of Life" talk. By February 2003, the stock of Geron, the company founded by Mike West, the company that claimed to "own" stem cells, whose slogan was "living cells are tomorrow's pharmaceuticals" with its vision of frozen packets of cells stacked up in hospital pharmacies and used like prescription medicines, had dropped to about $1.50. The company eviscerated its staff.

People were catching on to what Andrews already knew. The fact was, despite all the claims made on behalf of cloning and ES cells, biology had almost no idea how to make them useful to people. As time went on, and the topic of human cloning was discussed over and over again, even people who desperately wanted cloning to work, especially infertile couples who had been through the IVF wringer, said safety issues and the low success rate would stop them for now.

When you stopped to think about it logically, therapies using ES cells and cloning were a half-baked idea. For one thing, ES cells, by definition, cause cancer. That's the way biology tests for ES cells, which, when injected into a lab mouse, will form the very teratomas that Peter Andrews spent thirty years studying. Though teratomas are mainly benign, nobody was going to want one. As for custom-made organs, tissue engineers were certainly making progress on re-creating human parts in labs. Human bladders, human ureters, human ligaments were all being manufactured and the engineers were getting better at it virtually every day. Some engineered tissues, like ureters, were installed in people. An international project started in 2000 aimed to engineer a whole human heart. A Boston team made headlines by creating penises in a dish. But it's not like you could look forward to stitching on a Big Boss Man in the near future. The penises were for rabbits, they were pretty stubby, and while they worked well enough on actual bunnies to make baby bunnies, no human being was going to be thrilled.

Making a customized liver for your failing one with cloned ES cells would require taking cells from your body, putting them into an egg, getting the egg to make an embryo, harvesting the ES cells, turning the ES cells into liver and veins and all the other parts of a complex organ, laying the cells onto some sort of architecture, getting them to grow, and then installing the whole thing into your body. You might be dead by then.

Tissue engineering was probably going to work. It already did work for some applications and animal livers were, in fact, well on the way to being engineered. But the idea of a kitchen full of Wolf-gang Pucks whipping up replacement parts on demand was far, far from reality.

Even cloning was flunking the practicality test. PPL Therapeutics was faltering. Advanced Cell Technology was forced to sell its money-making cattle cloning division, Cyagra, to a wealthy ranching family conglomerate in Argentina. Infigen, the company that cloned "Gene," the first bull, was on life support, especially after losing a patent dispute to ACT. A Netherlands cloning outfit called Pharming went bankrupt. Nobody was making money by cloning. Sure, there was some interesting stuff going on, like goats that spit spider silk out of their teats, cows that pumped human proteins, and pigs whose organs could be used for human transplant, but this was research, not a business. Cloning to make living creatures was not going to go away, but it seemed destined to become a tiny niche.

So it seemed sadly symbolic that in February 2003 Dolly was put down by vets after developing a virally induced lung tumor, also called Jaagsiekte, a well-known disease in sheep. The disease had nothing to do with cloning, though television news made it seem as if it did, and her death elicited predictable dire warnings from anti-cloners. Once again, they had gotten her significance all wrong. Dolly had ushered in the bioutopian love affair with cloning and stem cell crazes. On Valentine's Day she ushered it out again.

MIKE WEST'S TROUBLES CAME, in part, courtesy of his own tunnel vision and the federal government, as if the conspiring Fates in the Greek drama he picked up while learning about ancient religions had sprung off an Attica stage.

West made a lot of news, but ACT had always been a small, struggling outfit occupying a suite of office space and a tiny lab, more of a research facility than a business. It subsisted, mainly, on income from its cattle cloning work and on federal grants meant to advance the therapeutic and economic potential of cells. These grants were funded by the NIH and by an agency within the Department of Commerce that encouraged the development of innovative technologies. But by 2001, West had used the company's powerful intellectual property position in the field of nuclear transfer to attract an all-star cast of cloners. In addition to Jose Cibelli, who had been there before West took over, West hired, among others, Teruhiko Wakayama, the man who first cloned mice and who had won an NIH grant, and Tanja Dominko, who was part of the lab that created the monkey with the "jellyfish gene" that so upset columnist Charles Krauthammer. She was on the verge of winning a grant from the Department of Commerce.

But soon after they arrived, West's mercurial management style, his constant changes in research direction, and a seemingly irresistible impulse to make waves drove the employees to explore jobs elsewhere. They figured that career security was more assured in a less controversial academic lab or another biotech business.

Those instincts proved correct. ACT's embryo work had drawn the wrath of the bio-Luddites, the Bush administration, and members of Congress, who wondered why the government should give any money to a company that was not only doing embryo experiments, but was using cloning technology. None of the government money was used for work on human embryos. It was kept strictly separate and used for other research. But the money clearly helped keep ACT alive as a going concern.

By February 2002, the government was subjecting ACT's scientists to frequent and thorough bookkeeping checks. Investigators were scouring the ACT lab, leading its scientists to suspect that the government was looking for any excuse to reneg on grants.

ACT did not know it, but Congressman Joe Pitts, a Pennsylvania Republican and an opponent of research like ACT's, had started making noise. He demanded an audit of the company. Within the grant-making agencies, a rumor began floating that the White House, which had so roundly condemned ACT and any research that made embryos or used cloning technology, risked embarrassment if it became known that the federal government was giving research money to ACT, public enemy number one among the bio-Luddites. The rumor that Bush asked Secretary of Commerce Donald Evans to make ACT go away became the accepted wisdom, and whether that was true or not, that's the message that came down to the agencies. "It's odd that [West] underestimates the fortitude of the forces arrayed against him," one government insider said.

During the late winter and early spring of 2002, the auditors combed ACT's corporate books, looking into a grant given to

Wakayama by the NIH. "They sent the office of inspector general [of the Department of Health and Human Services, overseer of the NIH] down here to hang us," West recalled. "They found no evidence that we used any money for human embryo work. Why would we? They knew the first day we had not. The first hour they were here, they were, like, 'Boy, this is gonna be a long thirty days.'"

While they failed to find any evidence that ACT had mingled funds, they did uncover the purchase of a microscope that had not been part of the original grant to Wakayama. West argued that he had received NIH approval for the scope purchase and that the auditors had called the NIH and confirmed this. (It is not unusual for a grant recipient to make such requests and for those requests to be approved since not every need can be foreseen in the original grant application.)

But West said the inspectors took their cues from Health and Human Services Secretary Tommy Thompson. "He told them they should nail us for fraud . . . they had Tommy Thompson on the phone in our conference room accusing our scientists of fraud. I was like, 'Come on, guys, this is over the top.' I thought it was a misuse of power . . . I was not proud of our country."

Pitts grandstanded on the microscope incident, insisting that ACT had misappropriated. "ACT is involved in unethical research that eighty to ninety percent of the American people want to see banned," Pitts said in a press release. "I was shocked to find out last year that millions of taxpayer dollars were supporting this company's research. That's why we asked for an audit . . . I suppose it's no surprise that their finances are as questionable as their ethics."

The federal government, Pitts told the *New York Times*, "should make a commitment to never again fund a company that does research the majority of Americans think is unethical."

ACT disintegrated. Wakayama went home to Japan. Cibelli left for the University of Michigan. Dominko departed amid a cloud of acrimony. Venture capitalists declined to extend more money to the company. West was rumored to be turning to Saul Kent for cash, though he softly denied it. At least one other antiaging organization West approached was leery of his reputation as an unfocused manager whose zeal for hitting splashy antiaging home runs sometimes caused him to lose sight of more immediate goals. West began responding to people who asked, "How's it going, Mike," by saying, "Well, we made payroll this week." (In March 2003, West did manage to attract some private investors he would not identify. "We're limping, but take us off life support," he said.)

"Are you paying a price for your agenda?" I asked West.

"Oh my gosh, yes," he said. Then he thought about Haseltine and the statements Haseltine has made about the prospects for a transubstantiated human future and for immortality. West's ideas were no more radical. HGS, though, had all that cash, and Haseltine was firmly ensconced in the science, political, and social firmaments. West sighed. "I am much more vulnerable than Bill Haseltine."

BILL HASELTINE DOES NOT APOLOGIZE. He declines contrition. He sees one way—forward—and one speed—full. Haseltine is a human being and he would prefer if people liked him, but he doesn't lose any sleep over whether they do or don't. He'd rather be right than be loved and at the Waldorf investors meeting, he refused to concede an inch. He knew he'd get questions like Duncan's, but to him, they were an opportunity to defend his case. He never forgot how much fun it was to go out on a limb and watch everybody else try to chop down the tree.

He had opened the meeting by ignoring the stock price collapse and all the bad news. "We hope you are enjoying your lunch and that you will enjoy the feast we hope to put before you in terms of our progress over the past year." He then mounted a metaphor-laden defense of the company. "We have built a sturdy ship to navigate across these turbulent seas." "Our engine for discovery is powerful. We have created what I regard as the most efficient means of converting ideas from genomics, monoclonal antibodies, and other sources into products for clinical trials." He then attacked the naysayers head on. "We have a record unequalled in the industry for efficiency. That is our legacy of genomics. People ask 'Where is the beef in genomics?' If you want to see where it is, it is in Human Genome Sciences today and, if people are clever, it will be in other companies tomorrow."

True, HGS had run into a rough patch of sea, but investors should buck up. Nobody said this was going to be easy. "Sometimes there are fair winds and sometimes foul. And I do not think you can describe the current market as fair wind." But through it all, he argued, Human Genome Sciences had stayed with genomics and would continue to do so because "this is our compass, how we navigate our ship."

Telling the world you could make new and better drugs on a massive scale by plucking gene signals from human cells, and then turning those signals into product was not popular at the time he first said it, and now the skepticism was being rewarded with each bit of sour news. But he was correct back in 1992 when he first met Venter, and now, ten years later, he was still right, he insisted, no matter what the peanut gallery might believe and no matter how much pleasure they took in the company's current doldrums. He reminded the crowd that he had said exactly what he would do when HGS began and that he had stuck to his word. He'd find genes and proteins and leverage that data by selling some of it to pay for HGS's own drug development. He pointed to Hal Werner, the Healthcare Ventures partner who was sitting in the ballroom audience, and then to Deeda Blair. They were there at the beginning, he said, and they'd back him up. Human Genome Sciences had done what it said it would do: it had discovered the majority of useful human genes.

Yes, yes, people didn't believe it since the genomes had been published and the sequencers said there were just 30,000 genes. But he was right and they were wrong. Who cared about genomes, anyway? Genes were what you wanted. "We discovered first more than 90 percent of human genes and we found them in a form that was useful and we found them by 1995. We know the rest of the world treats this discovery with a very skeptical eye but we have no reason to change our view. When we look at the genes we had in 1995 and we compare them to what the world knows, we still have 40 or 50 percent more genes than the rest of the world has described and we have been using those to great advantage."

Even in the midst of his ardent defense, though, Haseltine seemed almost bored. Human Genome Sciences had been up and running for almost ten years, and people were still asking the same questions. Didn't they get how this worked? This was biotech and in biotech, you always made big statements about breakthroughs and revolutions. So what if genomics was not the revolution Wallace Steinberg said it would be? What could you expect from a venture capitalist, especially a master salesman like Steinberg? Of course, Haseltine would be proven right about genomics, or at least right enough, and he took such vindication for granted. So put aside the PR flack. Genomics was just one more step, a tool, not a panacea. Aside from the technology created by Venter and the makers of the machines Venter used, which was a matter of scale, not break-through science, HGS was not all that different from Genentech or Biogen or Amgen. Biotech had been using genes to get protein drugs for twenty years. The top executives in HGS, like Rosen and drug development honcho David Stump, were veterans of the industry. Stump helped usher some of Genentech's drugs to market. Of course genomics would lead to new drugs, eventually, and people would make money. In fact, though Haseltine went to the office just about every day, the company practically ran itself these days. For him, the story had moved on from Human Genome Sciences. Why couldn't everybody else see the story he saw? He sometimes sounded as though he wanted to tell them all to just get over the minutiae of stock prices and drug trials, because he sure had. Regenerative med-icine was just getting off the ground, and beyond that, there would be more steps to take, and then more, and he was in a hurry to get moving. "I always try to look at what's next," he told me.

Most people failed to appreciate what had occurred over the pre-vious fifty years, but Haseltine, who had spent his entire adult life in the middle of it without losing sight of the big picture, did. Maury Fox and the rest of the early molecular biologists had sat around Cold Spring Harbor in their shorts oohing and ahhing over a grainy black-and-white film of cell division in corn. Doors opened that had been closed since the beginning of time and in the years since then, scientists had begun walking through those doors, making the world a much bigger place than it once was. Some of the doors would lead nowhere. "Science is hard," Haseltine said, "and medicine even harder." Human Genome Sciences would survive, and Haseltine would prove the naysayers wrong. But some biotech companies would fail. Maybe most would. Still, their very creation was testi-mony to the faith people had in the power of science to improve their lives. Steinberg stoked the faith, sure, but being the salesman he was he recognized just how desperately people wanted to live longer and to live better. A lot of people, and not just Extropians or transhumanists, couldn't wait and biotechnology became an act of faith. Like every religion, it promised more than it could deliver, faster than it could deliver. Like believers everywhere, the faithful kept right on believing, and even worked to manufacture the rapture that transhumanists, life extensionists, Extropians, digerati, the whole bio-techno-utopian clan had invented in their heads. But by 2002 faith in the biorapture was no longer a fringe curiosity. Deeda Blair was hardly an Extropian. In fact, the Extropians were in danger of becoming passé because the faith was now so ingrained in science itself, the ultimate skeptic, that biologists were walking bravely through the open doors and deciding that the human species really could give itself an enhanced existence and create its own immortality. The two sides, once leery of each other, were becoming indistinguishable. "There is no reason why we can't live forever," Haseltine said, "and there is no reason why we shouldn't."

10 THE RAPTURE RIDES IN A LIMO

SINCE THE AGE OF NINE, Haseltine believed he was on a mission, that he had been given gifts to complete that mission, and that therefore it was his duty to do so. Starting from that day in the Lansing airport, he decided that if he made a lot of money in the process, all the better, but rich or not, he had to speak the word. Not just about biology, but about everything. It was no accident that people like opera singer Beverly Sills, diplomat Richard Holbrooke, economist and former trade ambassador Laura Tyson were, at one time or another, on the Human Genome Sciences board, or that HGS's corporate reports and its corporate headquarters featured fine art from Roman-style mosaics to postmodern painting, an affectation that sometimes elicited snarky comments from business writers. Haseltine saw himself playing on the big stage of human well-being. He liked the status of being part of the Brookings Institution and the Trilateral Commission, of being chairman of the board of the National Health Museum. He liked giving talks to the Democratic Leadership Council, the organization that spawned Bill Clinton's run at the presidency, and presenting lectures and answering questions as part of panels in Davos, Gstaad, Aspen, Rome, Tokyo. All that was a long way from playing Daniel Boone in China Lake.

But though he reveled in being a big shot, and was famous within biotech for cruising in his limousine, the trappings were also a means to an end. "I have always had a compass that has guided me, determined not by the dynamics of science but by the dynamics of society and politics." Haseltine was in the happiness business and anything that deposited coins in the happiness bank was good, including stable government, great art, education, vibrant societies. Human beings have always tried to carve out better and better manifestations of all those things in order to up the happiness quotient, but now there was thinking abroad in the land that humans should exclude themselves from improvement. That didn't make any sense. As Haseltine wrote of bio-Luddite Francis Fukayama's book, *Our Posthuman Future*, "Achievements that I believe will one day be regarded as a golden moment in human accomplishment are viewed in this book through a narrow, fearful gaze."

What did people want just about more than anything else? To not die, to be healthy, and for their children to not die and to be healthy. That's why individuals tried to get rich, mainly—to avoid suffering, improve their health, live longer. To Haseltine, there was no particular glory in suffering, not if you could avoid it, and manipulating biology was an avenue to prevent it. He did not believe, as the nuns in my Catholic school asked us to believe, and as the bio-Luddites appeared to think, that suffering was a key to heaven, a way to find God. To Haseltine, human progress was about taking over from God to make human life better.

Haseltine seemed like a science outlier when he said things like this, but he wasn't. James Watson, for example, had once accused Robert Edwards of moving too fast with IVF, and in 1971 he wrote his famous essay warning of a coming age of human cloning. But by 2002, Watson was saying, well, so what? Why not clone? Why not design a disease-free kid? What was so awful about extending the human life span? "People say it would be terrible if we made all girls pretty. I think it would be great."

Watson could get away with such statements because he was Watson and famous for being a cranky old contrarian. Other biologists, both academic and corporate, began saying the same things at various points during the 1990s, though most who did had a habit of reaching across desks to flick off tape recorders before using words like *life span extension, immortality,* or *human enhancement.* They had spent so many years trying to sound rational, and not believing that such things were possible, that now that they had changed their minds, they still feared being regarded as kooks.

Haseltine had no such reticence. He used his relative fame, his connections to the Washington and New York power structures, his money, to push an agenda he saw as not only inevitable, but desirable. As a result, even scientists and biotech executives who did not like Haseltine had to admit that he had changed the rules and expectations. Rather than flicking off tape recorders, they began to say that of course biotech should pursue the rapture. They knew it wasn't here yet. Even West sometimes tried to sound cautious. But regardless of the fate of this or that biotechnology company, or even whole strains of scientific investigation like embryonic stem cells or cloning, the idea had caught on that Haldane and Schrodinger were right, that man not only could, but *should*, take control.

Despite all the bad news coming out of biotech, the belief in a paradisaical transubstantiated future is as strong as ever among those who have always believed and it has spread, thanks to the power of the biotech idea. The faith has been made enduring by almost a century of speculation. In the twenty-five years since the dawn of biotech, the speculation has turned into public relations, the clay of raw science shaped into companies. Venture capital has breathed life into the companies and biotech has enlisted everyone by going public and proselytizing the coming miracles. Never before in human history have people been so sure that science—not magic, not God—was about to mine the secrets of nature and turn them over to human beings. This is new. It's a religion in its own right and it is making converts.

FROM ACROSS THE TABLE, Jose Salgado points his fork at me accusingly and says, "What are you doing, Brian?"

At first I'm not sure what he means. I'm eating dinner deep in bowels of the Bellagio Hotel in Las Vegas, that's what I'm doing. It occurs to me that maybe I've loaded up on a few too many carbohydrates from the buffet, a Brobdinagian feast even by Vegas standards with piles of shrimp fried rice, ravioli, beef Wellington, cornbread—food from every continent. I've tried not to display favoritism. There is a world of starch on my plate. Starch makes sugar in your body. Sugar, as every life extensionist knows, is bad. When you burn it, the power plants in your cells send free radicals bouncing all over the place. The free radicals age the hell out of you. That's the theory, anyway.

Earlier in the day, a motivational speaker kicked off this weekend's Longevity Boot Camp by laying out twenty-one ways to live to
be one hundred and his number one way was "Stop eating breads."
They were part of the "three white poisons": sugar, salt, flour. Lean
protein, that's what you want, he said. Drink water until you pee a
river. Fruit and vegetables. Eat lots of fruit and vegetables. When
people follow that diet, he said, they come into clinics in wheel-
chairs and in two weeks, they are up and "dancing the night away."

I did not believe him and I gather Salgado did not believe him
either because when I look at his own payload of carbs I see that
Jose, a man twice my size, is apparently very, very hungry. I look
around at my fellow diners. Max and Natasha; Karlis Ullis, a doctor
famous in the life extension subculture; a biochemist named
Christopher Heward, and Stephen Coles—the same Stephen Coles
who touted co-enzyme Q10 to the life extensionists at the 1994 A4M
meeting, the man who hoped he could "come out of the closet" of
life extension—and his family have all come to the Bellagio and as I
look around, I am surprised to see every one of them indulging in the
three white poisons like crack addicts on vacation.

Seeing that his question confused me, Salgado tries to be more
specific.

"What are you doing to stop aging?" he asks, putting down his
fork to wait for my answer.

Salgado sounds like the paper folders the nuns used to hand out,
bluish homework keepers decorated with improbably clean-cut
youths. "Is God calling you," the covers asked in a none-too-subtle
stab at recruitment, "to be a priest, a brother, a sister?" Well, I reply
to Jose, I work out a little. I try to eat right. He laughs. No, no, no.
He doesn't mean just me, personally. How am I, he wants to know,
making a difference to mankind's greatest cause, the long war to
conquer aging and death?

I suspect that Salgado is about to put the bite on me, a natural
conclusion seeing as how some months before, in Max and
Natasha's living room, he told me about Project Life, an antiaging
foundation he was starting to fund research. I brace myself for the
coming pitch.

But he doesn't throw one. He just wants to know what I am
doing to encourage Bill Haseltine, and other scientists, to get seri-

ous about aging. I try to explain that Bill Haseltine doesn't really need any encouragement. He manages pretty well on his own. Besides, I tell Salgado, I'm not sure Haseltine listens to advice from others on this subject. If he did, he'd never mention it again because some people in his own company have suggested that he ought to stop talking about it. "I don't buy into some of the quacky wacky statements Bill makes," one employee told me, "like 'Death is defeat.'"

That's when Stephen Coles leans over to me and says, "Bill Haseltine is a hero."

How times have changed. Back in 1994, back in the old days of A4M, guys like Haseltine were the enemy. He was part of the science and medical establishment, a drug company executive no less, and so part of the giant conspiracy of big medicine working to keep life extension down so it could keep pumping people full of costly, patented medicines when everybody knew that a dash of St. John's Wort would set you right.

At first, I thought the Longevity Boot Camp would simply be a repeat of these old messages. The day began in the Golden Room of the much more down market Sahara hotel at the other end of the Strip. The motivational speaker kicked off. It seemed curious that a motivational speaker would be hired to fire up people to believe that they could live longer. The $495 each one had paid was pretty strong evidence that they were motivated already. "Decide not to be sick!" he shouted. This was standard, old-time, life extension rhetoric. But the rest of the speakers had a very different, and brand-new, message. Mainstream science was now our friend.

"We want to tell you how not to die of age-related diseases until the scientists and biotech entrepreneurs can deliver something to extend maximum life span." These were the words of Durk Pearson and Sandy Shaw. They walked up to the platform at the front of the room, sat in big wicker chairs opposite each other like life extension royalty granting an audience, and declared that it was no longer "better orange than dead." Now they used words like *apoptosis, genome, DNA methylation,* all molecular and cell biology buzzwords that were once foreign to life extension talks. Durk wasn't Jesus anymore. He was Moses leading the tribe to the gates. Beyond the gates, biotech awaited.

That the old methods were not working to stop aging was obvious. Pearson and Shaw did not look particularly well. Durk's hair had gone white and he carried a paunch on a frame curved with a pronounced stoop. Sandy was hard of hearing. They still advised using the Durk-n-Sandy products, but now the products were meant to help us hold on just a little longer. Marry everything life extensionists had learned over the years—the diets, the vitamins, the supplements—to the real science and you had a very good chance of making it over the hump.

Kimberly Frye, a fifty-six-year-old interior decorator from Jupiter, Florida, sat next to me as Durk and Sandy rattled off dozens of various preparations. She was getting confused. DNA? Apoptosis? She had been an antiaging consumer for the past seven or eight years. She had tried HGH for a while. She went to a few of these seminars. The work and money appeared to have paid off because Frye was pretty, fit, blonde, and looked at least a decade younger than fifty-six, but now she wasn't sure which way to turn. Should she just stop taking all the supplements and wait for the biotech payoff? Because if it really was coming soon, like Durk and Sandy and a few other speakers said it was, then maybe the products were a waste of time and money. Besides, if they really knew what they were talking about, with all their suggestions for what to inject, drink, swallow, and eat, then "why do these guys look like hell?"

The Longevity Boot Camp was the clearest statement yet that for years life extensionists were just whistling in the dark. They did the best they could. They tried to be scientific. But because real scientists thought aging was the third rail of biology, the life extensionists were on their own and the ratio of bogus to real always seemed tipped on the side of bogus. Now they admitted it.

Stephen Coles stood and delivered a talk he entitled "AntiAging Technology and Pseudo Science." He was still a believer in coenzyme Q10, he told the audience, but merely as a tool to steel your health. There was no scientifically proven antiaging medicine. To do much about aging, he said, we were going to have to intervene at the genetic level. Solving the puzzle of our decline, curing death, was going to be hard. But the solution was just about here. Coles believed it through and through. Forget about growth hormone. Forget about DMAE and SAM-E and DHEA and the scores of other crazy acronyms. They sounded science-y, but they were just cold cream on a sunburn. Real science was out there. Science was progressing so rapidly, that pretty soon everybody could kiss the offshore clinics, the strange supplements, the diets, goodbye. Archangels in lab coats were coming and they would usher us all into the great white light of biotechnology.

"We'll put new genetic programming into our adult stem cells." Coles put a lot of faith in stem cells of all kinds and in the idea of regenerative medicine. That's why he encouraged the life extensionists to step out of the shadows, stop talking among themselves at seminars like this, stop preaching to the choir. Our very lives were being threatened by the bio-Luddites. Thank goodness there are companies like Advanced Cell Technology and journals like Haseltine's *e-Biomed*, unafraid of the backlash. "Listen closely when you hear the name Michael West," he told the campers. Find ways to become politically active. Support the foundations sponsoring stem cell science. "Our side can still win the war. Remember, the grim reaper is still waiting for us."

Salgado picks up his fork and points it at me again. He's so frustrated, he says. Sure, lots of exciting developments have occurred in biology. Some scientists appear to have been converted to life extension. But it is still too slow. He's a middle-aged man. He's not really an Extropian, maybe not even a transhumanist. He just enjoys living, and wants to keep doing it. And no wonder. Here he is, an immigrant from Venezuela. He founded a company called Softni, an abbreviation for Software New Ideas, that makes products for adding captions to TV programs and movies so that Univision, for example, can broadcast U.S. shows into Latin America with Spanish subtitles. He's made a lot of money, gotten to know a few people in the Hollywood and Las Vegas show biz world—which is nice because it lets you jump the long line at the Bellagio buffet and get comped to boot. He's got a very cute girlfriend, a taste for wines and good food. Jose Salgado is the embodiment of why it's good to live in America. So why in the world would he want to die?

Coles chimes in. The old guard, like the founders of A4M, he says, are now seen within the antiaging movement as false prophets, sharp operators, growth hormone peddlers. The answer has been in molecular biology and in cells all along. Joshua Lederberg wrote a letter of recommendation to help Coles get into medical school and now Coles believed that, at last, the biology Lederberg helped to start, and the biotech industry that sprang from it, were stepping in and taking their rightful place as saviors.

Dinner is over. Most of the men head off into the night, to the first true transhuman pleasure palaces, the strip clubs of Vegas. There, as Natasha explains it, they'll admire the beauty of women who have already made the posthuman leap.

JOHN SPERLING AGREED that the rapture was coming. Good thing, too, because he was an old man and he wasn't looking forward to dying. Life extension, cloning, stem cells—Sperling wanted to do it all, and, like transhumanists, he thought it could all be done. But he had two ingredients transhumanists did not have. He had political connections that he was not only unafraid of using to antagonize the government, the bio-Luddites, the right-to-lifers, he reveled in the prospect. As far as he was concerned, the more people he rankled the more fun he had. Better still, John Sperling had a billion dollars.

He was the bio-Luddite nightmare.

Increasingly, just as the bio-Luddites feared, pop culture was becoming an enhanced experience. "Hey, everybody wants to look good," a photo retoucher told the *New York Times*. His work for major fashion magazines and celebrity photographers was all about image, and image, not MOSH, was where it was at. "To those who say too much retouching, I say you are bogus. This is the world that we're living in. Everything is glorified. I say live in your time."

Though he was old, Sperling lived in his time. Cosmetic surgery, tattoos, piercings, erections on demand, libido chargers, hair restorers, stay-awake drugs that had few or no side effects, Ritalin as a study aid. The bio-Luddites may have feared it all, but Sperling knew many people could not get enough. They were willing to pay for enhancements because they thought the augmentations made their lives better. They became bigger, stronger, sexier, smarter. Hardly anybody wanted to die. It didn't matter how much of this was perception or reality because in the biotech religion, perception created reality.

So Sperling decided to tap into both. In the process he may be presiding over the best example of the unification of the former fringe and the mainstream.

A long-term, ongoing science project had yet to be mounted by life extensionists because they treated longevity as a mission, not a business. For several years the life extension movement had tried to start a real-live clinic of their own, an operation that would use the latest, best, techniques and information that science could deliver. They'd be scientific about it. They wouldn't just gulp DHEA and betacarotene. They would conduct studies and do science and then follow wherever the science led. But the plans had gone nowhere. Anybody with the kinds of resources to spend on such a project either thought the whole idea was a crock or they had already given their money to mainstream researchers and institutions. Paul Glenn, another one of Florence Mahoney's life extension friends, who made a fortune in commodities trading, started the Glenn Foundation to distribute monies to promising research. Larry Ellison, a product of the Bay Area techno-fest, the founder of Oracle, and once the world's second richest man before the technology bubble burst in late 2000, established the Ellison Medical Foundation partly to fund scientists studying aging. (Ellison was sold on the promise of biotech to do great things. "If I were twenty one years old," he told *BusinessWeek* magazine, "I would go into biotechnology or genetic engineering.")

There had been freelance efforts within the life extensionist community to get this sort of thing off the ground, initiated by bioutopians who felt the rapture was at hand if only science would get on the ball. For example, David Kekich, a self-styled investment banker and financial planner, tried to take a page from Wally Steinberg's book. Kekich suffered a fluke spinal injury as a younger man and was now restricted to a wheelchair, but before his injury, Kekich was an athlete, a weight lifter, a devoted life extensionist. For someone who had been so interested in his health and prolonging his life, the injury was a very cruel cut. But he survived and even thrived thanks to a positive outlook and a determined vision that science would someday lift him out of that chair and extend his life to boot. So Kekich put his financial acumen to work by starting the Maximum Life Foundation and an associated business venture, TransVio Technology Ventures, which aimed to persuade business to use some intricate accounting as a way to fund a new venture capital effort that would funnel money to biotech start-ups that would not be afraid to call themselves antiaging. Kekich made his pitch to foundations that funded aging research, even to a representative of the National Institute on Aging, but they were dubious, wondering how money could appear, seemingly out of thin air in Kekich's complex accounting maze, to pay for longevity research. Kekich was sincere. He truly believed he had figured out a perfectly legal way to turn unused corporate assets into a stream of money for antiaging biotech, but the techniques were so convoluted his plan seemed doomed by the shadow of Enron.

Saul Kent's Life Extension Foundation was the model people like Kekich aspired to. Kent, the old cryonics warrior, and partner William Faloon started the foundation, its affiliated for-profit supplement retailer The Life Extension Buyers Club, and *Life Extension*, the magazine the foundation published, to generate income he could then use to finance scientific research. The Buyers Club grew into one of the world's biggest dietary supplement mail-order operations. *Life Extension* was an effective marketing and anti-FDA propaganda vehicle complete with pages of forms in the back of every issue that readers could use to purchase everything from Norwegian shark liver oil to something called mega GLA. Customers could even order up to twenty-nine different blood tests, the prices of which topped out at $450. They could also buy some of the classic life extensionist books like *The First Immortal* by James Halperin and *The Immortalist Manifesto* by Elixxir.

Sales royalties from all these products generated millions for the LEF. Between the royalties, membership dues, income from investments, and other proceeds, the foundation had amassed net assets of $17.6 million by the end of 2000, some of which Kent plowed back into research. He estimated that he spent between $2 and $3 million each year funding real scientists at the University of Wisconsin, the University of California at Riverside, and the University of Arkansas. His biggest grants, though, went to companies majority-owned by the foundation, like a small R&D company called 21st Century Medicine that experimented on new methods for freezing tissues, organs, and whole animals, like dogs.

Kent made a good living—his Buyers Club salary was over $400,000—but he was also serious about the research. By the turn of the century, cryonicists had admitted to themselves that people frozen before the late 1990s were goners. The old methods were just too crude. Alcor and the other cryonics outfits kept them frozen out of a sense of sacred duty, respect for the pioneers, and the slim chance that in some great technological leap, all the cells popped by ice crystals, all the cracked, mushy organs, could somehow be repaired. Who knew? Nanotech theorist Eric Drexler said that even if you were blown to smithereens, and there was nothing left but your big toe, nanotechnology could come to the rescue, millions of microscopically elfin construction workers rebuilding you. Yes, there would be the small matter of your profound amnesia. Your old brain, having become a slick on the sidewalk in the explosion, would take your memories, learning, and opinions with it and you'd have no idea who you were or how you wound up in a nanotech lab, but hey, medicine notches a save whenever the body recovers even if the patient has amnesia, so nanotech was perfectly justified in taking credit.

But when they weren't putting on a brave face, even the boundless optimism of the most ardent cryonicist gave way to the recognition that the prospects for even the more recently chilled were pretty iffy. So 21st Century Medicine rolled up its sleeves to improve cryonic preservation. Its two leaders even gained a measure of respect within the larger cryobiology community, especially for some of the work the two did to advance "vitrification," a technique they hoped would freeze heads so hard so fast, that the brains would turn to glass. But even Kent's efforts, directed as they were to cryonics, mainly, were falling short given the new age of cells and genes.

Sperling figured that if people were willing to undergo costly injections of fetal sheep cells in the 1930s, to take a medicine chest full of vitamins and supplements and exotic-sounding herbs in the 1980s, and visit Bergdorf Goodman and Saks 5th Avenue to pay hundreds for a face cream supposedly designed to complement their unique genetic makeup in 2002, they would also pay to have access to the very latest biotechnological tools. He was right.

For $6,000, customers came to Kronos, the clinic Sperling had founded in an upscale Scottsdale shopping mall. They met with a staff that included nurses, a doctor, a nutritionist, and an athletic trainer. They underwent a battery of physical, mental, and biomedical testing. They were served a gourmet lunch in a patient lounge paneled with blonde wood and decorated with cool Euro-style furniture. And then their blood was taken and sent to a state-of-the-art lab a few miles away across the desert floor where brand-new

machines that would be the envy of many biotech companies
churned and sampled and fractionated, testing for all kinds of "biomarkers," not just cholesterol or PSA levels, as in any normal blood
test a regular doctor might order, but more than a dozen parameters,
including all major hormones, levels of trace minerals, obscure proteins. Based on the results of this testing, a dietary, exercise, and
medication regimen was prescribed. Kronos then held the patient's
hand for a year in a series of personal and telephone consultations.

Kronos adopted a slogan: optimal health, the transhumanist
notion that "normal" was obsolete. Nobody is normal. Normal is old
medicine. Normal is the acceptance of the inevitable. We are all
either at greater or lesser risk for heart attacks, cancer, dementia,
high blood pressure, arthritis. Kronos tried to mediate whatever risks
its testing uncovered by aggressively prescribing drugs, hormones,
and supplements from its own compounding pharmacy in Las Vegas
and from Saul Kent's LEF. This was the "Kronos Way."

Customers had a taste for it. In 1999, Kronos earned about $2.5
million in gross revenue. That nearly tripled the following year. By
the end of 2002, Kronos was claiming to have taken in $20 million
and to have a client base of 8,000 people.

Kronos did not start out as an "optimal health" company. In 1999,
when Christopher Heward announced to the Extropians that someone had been found to finance the immortalist dream, Kronos was
most definitely "antiaging." It opened late that year to translate science into clinical practice in a self-conscious program of life extension. But soon after opening, the company dropped the phrase "antiaging," and it ceased talking about "life extension." Kronos wanted to
be taken seriously by scientists and "antiaging" had become so pol-
luted with hucksterism, so associated with "hormone mills," that scientists were leery of Kronos.

Kronos did sometimes prescribe growth hormone. Ever since the
days of the offshore clinics like El Dorado, doctors had discovered
that rich patients would pay a lot for it and, just as the bio-Luddites
feared, physicians in wealthy enclaves like Manhattan's upper east
side, Palm Beach, and La Jolla established "antiaging" clinics that
were not much more than hormone drive-thoughs. Magazines like
Palm Beach Illustrated carried full-page ads. "Grow Young Again"
promised an ad for the Karp Medical Centers. It pictured a nude,

svelte woman floating underwater like a perfect mermaid, the result, presumably, of growth hormone injections. The ad quoted the same old 1990 study that led Howard Turney to create El Dorado.

Even industry jumped on the hormone bandwagon. Smith-Kline Beecham sent a small salesforce to an A4M life extension meeting to promote its testosterone patch, a biotech product called Androderm. "AIM FOR THE PHYSIOLOGIC IDEAL," an Androderm brochure exhorted the A4M crowd. The cover showed a handsome, middle-aged man, carrying a lovely middle-aged woman in his arms with a kind of comic book thought bubble leading to a drawing of the heavily muscled, frontally nude, Greek god-man Heracles the archer about to unleash his arrow.

Hormones or not, and "optimal health" aside, nobody who was interested in life span extension could have any doubts about the ultimate Kronos goal, because by late 2002, Kronos and its parent holding company, Exeter Life Sciences, were busy trying to forge cooperation with Kent. But Kronos wanted to hire real scientists and too much talk about life extension, too many hormone prescriptions, could undermine credibility. That was the last thing Sperling wanted.

Few people outside of Phoenix knew much about John Sperling. Those who watched biotech closely, especially cloning, knew that Sperling had supplied the money used by cloning scientists at Texas A&M university to create Cc (Copy cat or Carbon copy—cloner humor), the world's first clone of a cat. Near the end of 2002, Sperling was also revealed to be the man who had created the Missyplicity Project, an effort to clone his dog, a Siberian husky mix named Missy. The Missyplicity Project was portrayed as just the sort of thing the bio-Luddites were worried about, a rich eccentric with too much money on his hands ordering up custom-made dogs. When Cc was born, even *Nature Biotechnology*, the establishment rooter for the biotech industry, editorialized against the project. "Cc was not created to advance medical knowledge or provide fundamental biological insights. She was created because there is a market among certain rich cat owners for resurrected animal companions." Actually, Sperling did not even want a cat. And there was much more to the Missy project than met the eye. Sperling was not an eccentric. He had a plan.

Sperling began life, no kidding, in a log cabin in the Missouri Ozarks. His mother was a Bible-quoting fundamentalist Christian racist, his father a feckless sometime railroad employee who rarely worked, drank heavily, and beat his son. His father's death, when John was still a boy, was the happiest day of John Sperling's life.

As a child, Sperling attended school sporadically, became a loner, and lived in fear of physical pain. At age seven, he had to have surgery to drain septic fluid from a lung but was drugged only with a local anesthetic. The pain was so excruciating that during the operation, he begged to be allowed to die. He never forgot the torment.

Following his father's death, Sperling and his mother moved to Oregon, where he attended high school. One day, Sperling was walking to meet another high school student, a girl he had begun to date. His body was telling him that he must, as soon as possible, have glorious, erotic sex with the girl. His mind, however, knew for certain that, should his body win out in the eternal debate of young Christian manhood, he would commit a grave sin. His mother had warned him that sex outside marriage was a one-way ticket to hell on the express bus. Poor John realized there was no way he could resist seducing his date, so he stopped walking and begged God to strike him dead—right then—to save him from eternal damnation. God declined the offer. Sperling rejected God, religion, and the fear of hell on the spot and never looked back.

Sperling managed to finish high school in 1939 and joined the merchant marine in an effort to escape Depression-era unemployment. He saw much of the world in his two years aboard ship, but he also learned a great deal from his fellow crewmates, some of whom were educated men who had signed on to escape the Depression, too. They introduced him to books and though Sperling had barely read a single book up to the time he became a sailor, he now devoured them, a cafeteria style of intellectual pursuit that stayed with him for the rest of his life. In casual conversation, he would veer from the thinking of philosopher John Searle to quantum physics in a single paragraph.

After his stint aboard ship, he attended Reed College in Portland, Jean Haseltine's alma mater. He took flight training during World War II, but was never sent overseas—a lucky break that enabled him to finish Reed, then go to graduate school at UC Berke-

ley, where he specialized in English history, philosophy, and economics and where he was introduced to dope, cheap wine, and the first
stirrings of the sexual revolution. He was awarded a three-year stu-
dentship to King's College, Cambridge, where, between love affairs
and a messy domestic life, he took his Ph.D.

Perhaps because of his experiences with his father, John Sperling
developed a visceral denial of authority. When he went to teach at
Ohio State University, he despised the department chairman. Later,
at San Jose State University, where he spent twelve years on the fac-
ulty, he organized a union strike, defied the administration and many
students by mounting an antipollution protest featuring the burial of
a new yellow Ford Maverick, and helped organize the local branch of
the ACLU.

Sperling's most defiant act, however, was the creation of a for-
profit educational center called the Institute for Professional Development, and then another for-profit venture called the University of
Phoenix. The idea behind both was to help working adults further
their education and earn college degrees by taking classes between
working hours. Almost from the beginning the schools were dubbed
"diploma mills," but they were very popular with the people who
took classes there because Sperling had successfully mined the
American desire for a second chance. He gave his customers access
to the kind of reinvention that was typically available only in standard, and much less convenient, institutions. The education establishment, the college accrediting agencies, fought him. He was even
investigated by the FBI, which accused Sperling of bribing public
officials in California. But the University of Phoenix spread across
the nation, including an outpost in Rockville, Maryland, on the first
floor of the Human Genome Sciences headquarters building.

In 1994, the holding company founded by Sperling, Apollo Group,
went public. Twenty years before, in 1974, he had started what became
the University of Phoenix with $26,000 in savings. The day after the
Apollo Group IPO, Sperling, at age seventy-three, was a billionaire.

When you are a defiant seventy-three-year-old billionaire who
started life peeing in an Ozarks outhouse, whose biography makes
Horatio Alger stories sound anemic, you are not only unafraid to
pick a fight, you can afford to do it. So when Sperling looked around
for something to do with his billion dollars, he decided to wage war

on the war on drugs. In 1996, Sperling, along with financier George Soros and insurance tycoon Peter Lewis, mounted ballot initiatives in Arizona and California to permit medical uses of marijuana and lessen penalties for some drug crimes. Both passed, much to the consternation of Bill Clinton's antidrug czar, General Barry McCaffery, the conservative Arizona congressional delegation, and the state's law enforcement community. In 1998, the trio funded more initiatives around the country and they passed, too.

Sperling was having fun. He was tweaking the power structure. He had homes in San Francisco and Phoenix, a private jet, a girlfriend. Too bad it all happened so late in life, he lamented. So few years to roll around in it all. But then Sperling happened to read a *New York Times* article about a worm. The name of the worm was *C. elegans*.

On August 15, 1997, a team of worm people in the lab of Gary Ruvkun at the Massachusetts General Hospital's department of molecular biology published a paper in *Science* about daf-2 worms. "Decreased DAF-2 signaling also causes an increase in life-span," read the paper's abstract. "Life-span regulation by insulin-like metabolic control is analogous to mammalian longevity enhancement induced by caloric restriction, suggesting a general link between metabolism, diapause, and longevity."

The only tool that had ever worked to extend the life span of a mammal had been calorie restriction (CR), which is why a few diehard life extensionists suffered through CR diets. Now Ruvkun's lab was saying that the action of genes might explain why CR worked and if that was true, maybe some sort of genetic manipulation could mimic the effects of calorie restriction. Sperling read the report on Ruvkun's paper in the *Times*, gave it to his cardiologist, and hired the doctor to research the entire field of the science of aging.

Had the cardiologist been given the task just seven years before, there would not have been much to report. But in 1990, Tom Johnson had discovered the age-1 mutation in *C. elegans*, and the movement of biologists into aging began, as a trickle at first, with people like Ruvkun, Kenyon, and Lithgow, and then a steady stream. The longest-lived human on record, Frenchwoman Jeanne Calment, was 122 years old when she died, but now it looked as if human beings could live far longer if you adjusted genes in just the right ways.

There was more. Richard Cutler, the National Institute on Aging scientist who spoke to A4M back in 1994, had long advocated a theory of aging based on free radical damage, a theory that had been around since the 1950s but had drifted in and out of acceptance. The free radical theory of aging said that toxic molecules released during the production of energy bounced around inside cells and damaged them like golf balls flying around in a Steuben glass boutique. Damaged cells, so went the theory, led to run down tissues, and that led to the wear and tear of aging that eventually led to death. The animal experiments seemed to show some connection between free radicals, metabolism, insulin, and calories. There was the Rudman growth hormone study that had so captivated the monied class. Growth hormone was part of the insulin-making system. It all became very intriguing.

As Sperling's doctor discovered, there were a lot of loose ends, but the loose ends might all be connected, somehow, if only the research would be done. When Sperling heard the report, he was convinced. Just as he had philosophized about education, that it should not belong to a ruling clique but should be as democratic as possible and open to capitalists, Sperling thought science should not belong to a clique of academics, government agencies, and the corporate scientists who shuttled between the other two. Anybody ought to be able to jump in, just like he jumped into education to create his own university, to attempt the right combination of therapies, liberal use of new science, and continued research so that aging might be conquered.

In 1998, Sperling hired a young health care executive and placed him in charge of a brand-new shell company called Exeter Life Sciences, named for the street where Sperling lives. Then Exeter called on the Los Angeles Gerontology Research Group, and its mainstay doctor, Stephen Coles. Coles plugged Exeter into the southern California life extensionist and Extropian networks. It was there that Exeter hoped to find somebody to direct an antiaging clinic. The ideal candidate would be a scientist who knew a lot about testing for many different kinds of substances in the body, like hormone levels.

Christopher Heward was perfect. A card-carrying Libertarian who resembles a leaner G. Gordon Liddy, Heward believed that people ought to be free to inject or swallow whatever they wanted as long as they understood the risks. Heward had made his name researching, melanocyte stimulating hormone (MSH), a hormone that triggers protective skin tanning. The idea was to use MSH as a cancer diagnostic to home in on skin cancer. Heward tried the stuff himself and found that not only did he catch a quick tan, he also developed a marathon erection that, unlike harmful priapism, wasn't painful in the least. Well now, thought Heward, there ought to be a good business selling a substance that could turn you into a studly George Hamilton. He recruited an Asian businessman to try to sell the idea in Japan, or Korea, but the business never flew, partly because when the businessman injected himself, he became violently ill; he and his erection spent the better part of a day bent over a toilet bowl.

Following the publication of a popular life extension book entitled *The Melatonin Miracle*, Heward and a partner began synthesizing melatonin, a brain hormone, creating some of the purest available for the dietary supplement industry. Along the way, Heward became a dedicated life extensionist. He began presenting talks to Coles's group, and to the Extropians, and, at one meeting, met Greg Stock, who had begun his advocacy of human germline modification. He and Stock formed a tiny company called Biometrics to track the age-related declines in so-called biomarkers like growth hormone. By late 2002, Heward was not just running Kronos, he was gatekeeper of science for Exeter and what was becoming John Sperling's mini biotech empire.

As soon as Kronos was established, Exeter created the Kronos Longevity Research Institute and installed the NIA's Richard Cutler as its head. Cutler recruited fellow NIA scientist S. Mitchell Harman, an endocrinologist who had participated in some groundbreaking studies on hormone replacement therapy (and whose wife had grown up with the Haseltines at China Lake). Cutler was a longtime acquaintance of Florence Mahoney and he and Harman were among the NIA staff who believed in her original antiaging vision for the institute. Now, with Sperling's money and promises of academic freedom to pursue longevity research, they had a chance to fulfill her ambition of using science to solve the problem of old age.

Cutler was thrilled. When the NIA was first formed, the other institutes laughed, so the institute played it safe and did not focus on the molecular process of aging, choosing instead to ameliorate the effects, insisting that it wanted to add "life to our years, not years to our lives." "My complaint is, why not have both?!" Cutler argued. "I do not think you can do one without the other. There is this thinking that it is bad to really interfere and increase life span, but we've never had a Manhattan Project whose aim and direction was to go after this thing called aging. That was Florence's desire."

Now things were different. Sperling was willing to spend real money. "I got approved for a research project at a million dollars per year to investigate longevity genes. I have never been able to get that type of money [from the government], not in seventeen, eighteen years. It was too far out." Cutler was most excited about a project called "Mighty Mouse," a prolongevity experiment that would try to engineer a mouse that lived dramatically longer than normal mice.

Haseltine was right, Harman said. There would be practical immortality. Meanwhile, it was time to start applying the latest life extension technologies to humans even while research continued. Give people a little more time, a little more health, "until the real answers to the question of what we can do to delay or reverse aging are available. They are not yet available, but they are back there in the gene lab, the rat lab and they are coming along. In the meantime, we are getting the infrastructure in place, so that when it emerges from the lab, we'll have the people, the techniques, the equipment."

Sperling had complete near and long-term visions that also encompassed Haseltine's pet idea, regenerative medicine, and the eventual enhancement of the human species so we could be smarter, disease-free, plain old happier. This was the real purpose behind Missyplicity. Cloning Missy would have been nice ("I wanted to clone my dog! Dolly's a goddamn sheep, why not a beloved dog?"), but the Missy project was "just what Aristotle called the 'efficient cause.' It would lead to an increase in the knowledge of science and the reproductive process, and we learned an enormous amount." Sperling used Missy to help create an infrastructure and a potential revenue stream to finance regenerative medicine and animal cloning. He set up a company called Genetic Savings and Clone, and installed his girlfriend's son to run it. The idea was to create a gene bank where pet lovers could store cells from their animals for use in later cloning experiments. The cash from fees could then pay for research.

Texas A&M's cloners never managed to clone Missy, a failure Sperling blamed on the scientists and what he claimed was their lack of focus. There was also a contretemps over Cc. Sperling claimed he never wanted a cat, never approved work on a cat. "They should not have been doing it on my money!"

What's more, Cc turned out to be a PR problem for Genetic Savings and Clone because of the small matter of Cc's appearance. While Cc satisfied a scientific purpose, she was lousy for business because she was the clone of a calico cat. Calicos have varying coat colorings, a trait that depends as much upon environmental cues within a mother cat's womb as it does on genes. So Cc wound up looking almost nothing like the cat she was cloned from, a bad omen if you're trying to convince people you can give them a reasonable facsimile of a beloved pet.

Still, Genetic Savings and Clone managed to attract paying clients. Meanwhile, Sperling bought out a tiny animal genetics company in Austin called GenomicFX that had been started by a couple of Texas A&M scientists. It was renamed Viagen. Missy's cells were sent there and Viagen began advertising for cloning specialists on e-mail newsletters circulated among embryo and IVF scientists.

Next, Exeter Life Sciences went fishing for rights to cloning technology. It nearly bought PPL Therapeutics, the company that collaborated with Wilmut and the Roslin Institute in the cloning of Dolly, in order to obtain license to the cloning patents. The PPL deal fell through, but Sperling was undaunted. He nearly bought West's Advanced Cell Technology to acquire licenses to its cloning techniques, but decided to obtain new cloning methods from a company started by ACT's original scientist-founder, James Robl. Then he set his sights on Prolinia, an agricultural cloning company founded by Steve Stice, a longtime ag cloner and veteran of Advanced Cell Technology. Prolinia held licenses to use the Roslin technology.

Sperling was trying to link cloning to Viagen's genetics expertise, and then to stem cell researchers in Asia. Though the near-term goal was agriculture, the ultimate ambition was to manipulate human cells. Sperling wasn't worried that the bio-Luddites would prevail and the United States would ban therapeutic cloning for stem cells. Not only would he enjoy the battle, he already had another option. Kronos had set up an "optimal health" clinic in Singapore. Exeter had very good relations with its local bank there. The bank had very good relations with local scientists and with scientists in China who were only too happy to use cloning technology.

While these maneuvers were costing a lot of money, about $50 million as of early 2003—more money than the life extension movement could conjure in its fondest wet dream—it was pocket change to Sperling. He was pretty sure any one of the companies could become a billion-dollar enterprise and all together, the worst he could do, he reckoned, was wind up with a set of sister enterprises worth half a billion. He was eighty-two years old, though he did not look it. He joked that he could fall over dead at any moment. Still, he figured that his last-minute dive into the world of biotechnology might well prove his greatest legacy. Just to be sure, he had set up a foundation. Upon his death, his billion would go there to be administered by Heward, the Exeter CEO, and a longtime aide from the early days of the University of Phoenix. The work would go on.

JUST AS HDTV and flat screens and broadband Internet connections and stainless steel kitchens changed the way people in much of the developed world viewed a normal house, the biotech religion changed the definition of a normal human just as the bio-Luddites feared it would. Kronos was only slightly ahead of the curve. There was no normal, not in the religion of biotechnology. The already blurry lines between disease and nondisease were completely erased. Everybody knew that if you had a heart attack, you had a diseased heart. But what did you have if you had high cholesterol? Everybody knew what Alzheimer's could do and if you had it, you had a disease. But what if you had a crummy memory? What if a pill or a gene therapy could give you a better one? True believers ignored the pop in the biotech bubble and stayed focused on the faith.

"Advances in biotech have prompted a new definition of disease," concluded a report by The Institute for the Future, to include "any outward expression of human aging. Using this as a definition, changes like wrinkles, menopause, and memory loss are all targets for intervention/prevention."

It wasn't just futurists who saw the coming rapture. The American Academy for the Advancement of Science, the publisher of the journal *Science*, and normally a very conservative group, issued a report of its own on the possibility of altering ourselves and giving our children enhancements by changing the germline. They were opposed, but, they said, it would soon be here. "Greater knowledge of genetics is also making it possible to contemplate genetic interventions not only to treat or eliminate diseases but also to 'enhance' normal human characteristics beyond what is necessary to sustain or restore good health. Examples would be efforts to improve height or intelligence or to intervene to change certain characteristics, such as the color of one's eyes or hair."

Though the science lagged, the concept of enhancement had become so mainstream that the National Science Foundation and the Department of Commerce held a workshop in December 2001 to push programs that would hurdle the obstacles. Entitled "Converging Technologies to Improve Human Performance," the report issued by the workshop participants recommended "a national R&D priority area on converging technologies focused on enhancing human performance." The combined technologies included "nano, bio, info, cogno" (nanotechnology, biotechnology, information science like computers, and brain "cognition" science).

"It is time to rekindle the spirit of the Renaissance," the document declared. Just as the Renaissance established the primacy of science over myth and combined the arts and sciences to create a new Western world, "essential features of the new society will be established in the novel, unified structure defined by converging technologies to improve human performance."

The transhumanists and those who thought like transhumanists had spent the past twenty-five years so far on the outside that when they looked back and took stock of where they were now, they could barely believe how close to the center they had come. They were thrilled that science was starting to accept their "memes." Sometimes it seemed like, pretty soon, MOSH really would be old-fashioned and Donna's dream would come true. A few people were already equipped with computer jacks in their skulls as a treatment for severe tinnitus. In early 2003, scientists at the University of Southern California announced they had created an "artificial hippocampus," the part of the brain that stores memories. Scientists created nano computers using biomolecules so small that one trillion computers could team up in a volume of a tenth of a milliliter to perform a billion operations per second with greater than 99.8 percent accuracy. "This study may lead to future computers that can operate within the human body," boasted the announcement of that experiment. Cognitive scientists trained a monkey to play a video game, implanted tiny electrodes into the monkey's brain, then trained it to play the game by thinking about it. The monkey was able to move the game's cursor by his thoughts alone. Brain cells from rats were placed on silicon chips and used to control robots, the first creation of a hybrid biological/mechanical cyborg.

In just ten or twenty years, said the "Converging Technologies" report, human brains will communicate directly with machines via superfast broadband connections. We'll be smarter, we'll learn more quickly. Our bodies will be "more durable, healthy, energetic, easier to repair, and resistant to many kinds of stress, biological threat, and aging processes . . . Engineers, artists, architects, and designers will experience tremendously expanded creative abilities, both with a variety of new tools and through improved understanding of the wellsprings of human creativity. The ability to control the genetics of humans, animals, and agricultural plants will greatly benefit human welfare."

It wasn't just smarter, healthier, long-lived people, but a whole new blissful, smiling age, a heaven on earth even Edward Bellamy could not have imagined. "If we make the correct decisions and investments today, any of these visions could be achieved within twenty years' time. Moving forward simultaneously along many of these paths could achieve a golden age that would be an epochal turning point in human history . . . Technological convergence could become the framework for human convergence. The twenty-first century could end in world peace, universal prosperity, and evolution to a higher level of compassion and accomplishment. It is hard to find the right metaphor to see a century into the future, but it may be that humanity would become like a single, transcendent nervous system, an interconnected 'brain' based in new core pathways of society." The earth-brain philosophy of Pierre Teilhard de Chardin, transhumanist hero, was now mainstream stuff.

Scientists were not just trying to enhance human powers by merging them with machines. There were ways to shut down genes or turn them on like so many light switches using molecules borrowed from moths, and another molecule called "zinc fingers." The Defense

Advanced Research Projects Agency (DARPA), the people who brought you the forerunner of the Internet, had an ongoing program "to explore augmenting human performance to increase the lethality and effectiveness of the warfighter by providing for super physiological and cognitive capabilities." The agency was researching ways to amp up human metabolism and to slow it down so far that humans could exist in near stasis, a lot like Heinlein's "cold sleep." Already there was promising research that could lead to the creation of drugs "to prevent the effects of sleep deprivation over an extended period of time, nominally set at up to seven days. At the end of the program we expect several candidate drugs that alone, or in combination, extend the performance envelope." A drug called modafinil was already on the market, in fact. It had been approved for narcoleptics, but it was being snapped up by shift workers, rave goers, cramming students, and they were finding that it worked with virtually no side effects.

DARPA also formally endorsed regeneration. In March 2003, the Defense Department decided that the idea was real enough to explore turning it into a program by inviting a small group of scientists—including Dominko—to a meeting in San Diego. The hope, according to a consultant's presentation to DARPA, was that people could recapture a talent human beings had lost to evolution, to "re-invent this 'lost art'" of amphibian regeneration. "Discovering ways to tap into these 'primitive' regenerative processes could help wounded soldiers contain traumatic loss, and perhaps even re-enter battle with new limbs." They weren't kidding. In the hotel bar after the first day of discussion, three scientists insisted that within three years they would be able to regrow a human finger. DARPA gave the project a Heinleinian name: "Regenesis."

Enhanced life had been created in laboratories. Muscular mice were genetically engineered. Smart "Doogie" mice were able to learn and retain more information than their unenhanced brethren. In one series of experiments, scientists at Merck created mice with better memories, taught them how to swim a water course, switched the gene off, and tried to teach them another one. With the gene off, the mice flunked.

Life was not just being enhanced, it was being created. Scientists at the Scripps Research Institute in San Diego invented a whole new species of E. coli bacteria by giving it a new genetic code. All known life on earth uses twenty amino acids. These scientists made a twenty-first, installed a synthetic gene to make it, and then watched as the new gene gave the *E. coli* resistance to a poison that killed regular *E. coli*. The next step, they said, was to make a mouse or a monkey with a new, synthetic genetic code.

There was no particular scientific reason why a scientist could not do any of these things to people. It was scientifically possible to create new generations of genetically modified babies just as the bio-Luddites warned. In fact, there were at least three ways to do it. You did not even have to have an embryo in a dish. For example, the cells that give rise to sperm cells could be genetically modified and then reimplanted into the testes. From then on, they would make genetically modified sperm. Good old-fashioned sex could deliver the New Age payload.

If it all sounded just too sci-fi, if anybody doubted that any of this was really possible, that a scientist could, say, make a genetically engineered superathlete, the World Anti-Doping Agency, the body that assists the International Olympic Committee in monitoring the use of performance-enhancing drugs, was pretty sure that athletes were about to start genetically modifying themselves. They had already done it, in a way. They were injecting growth hormone and a hormone called EPO to give themselves increased strength, power, and endurance. Growth hormone and EPO are the products of genes. The fact that they came out of big tanks full of *E. coli* and then had to be shot into a thigh didn't really matter so much. Biotechnology created enhancement.

Of course, VCs thought all this enhancement made a great story, too. Healthcare Ventures and several other gold-plated investment companies recruited a 2000 Nobel Prize winner named Eric Kandel to form a company called Memory Pharmaceuticals.

Developments like these, just as they had since FM 2030 first started collecting science journals, kept the faith alive.

THE TRUE BELIEVERS DENY their faith is a religion but it is and, like all religions, there are opponents who condemn it. Some argue it is a heresy upon the dogma of human nature. Others appeal to traditional religions and say the new one defies the will of God. The bio-Luddites insist that it threatens the natural order of human biology.

But there never was an orderly human biology. Biology is jazz. If it wasn't the transhuman future would already be here because biology would be so predictable we'd know all the rules and how to use them. But biology is complicated, as biotech continually discovers. It can, and does, go off on wild Louis Armstrong riffs, forgetting the melody line entirely. The very rise of the human species may have come as the result of a freakish fusing of two genes, an accident twenty-five million years ago. Much of the human genome is borrowed riffs from bacteria and viruses. In fact, there is no such thing as The Human Genome. Everybody has their own genome and, in the case of the Celera genome project, the human genome turned out to belong to Craig Venter. "The gene pool," Theodore Friedmann said, "is a heuristic fiction."

Even one person's genome is a jazz score. Mutations can be scribbled by, say, a virus that leaves its note and gives you cancer, or maybe does nothing at all but hitch a free ride. Everywhere you look there are soloist riffs being played. If one or more genes don't do precisely what they are supposed to do at precisely the right times, babies can be "ambiguous." Some people are born, literally, with the genes of two people. They are chimeras who merged in the womb and might be hermaphrodites with a mix-n-match set of genitals. Others are "mosaic," with unpredictable genetic instructions. Mrs. J'Noel Gardiner, a Kansas housewife, was born with male anatomy but was emotionally and mentally female. She had operations to give her the right equipment, married, but then found herself locked in legal combat with the son of her husband over her husband's estate. "We're talking Brave New World here," the son's lawyer said. "It's an illusion, it's an image she's trying to project," the son said of his stepmother's femaleness. "But it doesn't change the laws of God."

There were no laws of God regarding what biology could do. That was Haldane's message amid the disillusion of the post–World War I world. All bets were off. There were no biological tethers, at least no permanent ones. It was time to make a new reality to transcend the old, broken notions.

That has been the animating idea of biotechnology. A monkey with a jellyfish gene was not a "jonkeyfish," it was just a monkey whose DNA happened to glow green. But "Let's take monkeys" Douglas Melton suggested to me, "and take human ES cells and put them in a monkey blastula [embryo]. So here's the question: What portion of the monkey's brain and vocal cords do I need to have composed of human cells to allow the monkey to speak? This is an extremely legitimate scientific question . . . Suppose I discover you only need this portion of the brain up here to give the monkey speech," he continued, indicating a small area near his forehead. "This is a profoundly interesting experiment in my opinion. If you and I were to walk into my lab right now and the monkey would say 'Oh, Doug, so this is Brian?' It would chill us, right? That would really say, What is it about being human?"

Indeed. What is it? Biology is jazz. It is flexible, it breaks the rules, it allows for any number of improvisations, and, like jazz, it is subversive. That's what scares the bio-Luddites. "Make no mistake about it," the great ecosystem biologist E. O. Wilson told the *Wall Street Journal*. "The expansion of human knowledge with science and technology, especially neuroscience, genetics and evolution, renders traditional religious belief less and less tenable." If biotech keeps pushing to improvise the human musical score, people might be forced to confront the idea that we've been making all this up— what it means to be human, human culture, human society—as we go along.

That's why the bio-Luddites are wrong and it's why the bioutopians are wrong. It's why Haseltine is right. Even if you made Wally Steinberg God, and William Haseltine Jesus, and Deeda Blair the Virgin Mary, and all the bioutopians the Apostles, even if they made all the diseases go away and imparted immortality and introduced a full-option LX model human, there would still be no posthuman future. A thousand years from now, people will wish they felt as good at 520 as they did at 350. They'll wish their kids would have listened and not ruined their lives by becoming artists instead of lawyers specializing in interplanetary torts. People will still hate each other. They'll still think sex feels good, no matter what form it takes. Our descendants will all still be human. No matter how long humans live, how magnificent we become, we'll just be longer-lived, more magnificent people because we will still be asking—and still unable to answer—Sally Fox's question: "Why?"

Humanity is not a set of genes. It is a concept. Without the concept, we are, literally, biological machines created by accident out of the cosmic jam session of the Big Bang. But we became transhuman a long time ago when we gave ourselves the most profound augmentation possible, when we decided to make up our own definition of human and declared that we were more than a machine. That definition can be extremely broad. It does not have to be moored. It can be whatever we say it can be, because we're the ones who make it up. So if I walked into Doug Melton's lab and a monkey said, "Hello, Brian. How about a little poker because I feel lucky today, you look like a loser, and my wife's been busting my chops to fix the leaky toilet so who wants to go home to that?" I'd say I was playing poker with a human being who looked a helluva lot like a monkey.

But then I like jazz.

Adherents of the biotech religion argue that evolution has spent the past five million years enhancing the human species. What's wrong, they want to know, with taking a hand? If you don't believe in God, and you don't believe in heaven, then what are you supposed to do when you stare into James Frazer's "darkness of the grave"? "If you are not religious," Richard Cutler said, "then you see biotech as a hope." Otherwise, your only choice is to sit in a French café "feeling depressed and drinking absinthe." The goal is not just to give people longer lives, but to change the civilization. There are signs of hope this will happen, Cutler said, that people will recognize the power to be gained from letting go of the biological tether and realizing that if we are indeed making all this up as we go along, life, culture, civilization is all the more fragile and precious because they are not ordained. But, Cutler's colleague Mitch Harman said, the religion is still tenuous. "There is an enlightened 3 percent but we still have fundamentalists."

"We need to keep this going," Cutler interjected. "We need to hope this process will take place. I think the problem still is the basic philosophies and religions that dominated over the past several thousand years. There is going to be a big change coming."

"They won't go without a struggle," Harman cautioned. But Harman was positive the bio-Luddites would be overtaken by science. "We'll blow right by those people."

Florence Haseltine agreed. "There are people who want to control everything," she said. "There are people who want to control when people have sex, so the fact you want to control when people live or die does not surprise me. But it is going to be out of their hands, out of the politicians' hands. The science will leapfrog over them. Say you ban stem cells. And they find the body's stem cells do the same thing. Then you ban not dying and people get around it some other way. Who knows?"

"We need to do an unlikely thing," green bio-Luddite Bill Mc-Kibben has written. "We need to survey the world we now inhabit and proclaim it good. Good enough."

Not bloody likely. Humans have never done that. Science and the people who want to use science for whatever end they think will make them happy, will prevail. When IVF first came to the United States, Florence Haseltine was one of its earliest practitioners. At the time, IVF was so new there was no such thing as a success rate. Because the government refused to fund any research, IVF clinicians made up the science as they went along. Patients didn't care. "The early patients knew it was experimental," she said. "And they paid for it themselves. We were running off bootstrapping." She wasn't amused by the great faith people now placed in biotechnology. "I think they could have a lot more faith in it . . . there will be a lot of surprises. We have no idea what is gonna happen."

Even scientists who consider themselves open to genetic enhancement are sometimes shocked at how eagerly their students embrace the concept. "Every generation has a lower threshold," one told me. "Expectations change fast, especially if people think they will get something better at the other end." As Thomas Jefferson wrote, "the earth belongs, in usufruct, to the living." Someday, people will use genetic engineering on themselves and someday we will live far longer than 122 years. Much of the bioutopian creed is going to come true because imagination, a most human trait, will not be suppressed.

Donna will continue to use every tool in the biomedical kit in an effort to achieve gynoidom. The Extropians will continue to extol the ultrahuman Nirvana. Most people though, will skip the prix fixe utopian future. They'll go a la carte. Right now, people would like their cancer cured. They'd like to feel good at eighty. They wish their kids would not carry time bomb genes. If a leg went missing, they'd like to have it back, thank you very much. One of the bio-Luddite arguments against the expanding use of biotechnology is that diseases like manic-depression have helped produce great art from people like Beethoven, which is nice for the rest of us, but not so great for Beethoven. Most people do not want to suffer and they do not want to be martyrs to "human nature." If a pill could give you a better memory, would you turn it down? Who would refuse to regenerate their organs?

For the true bioutopians, the transhumanists, the Extropians, the life extensionists, for people like Donna who don't have a label, the faith remains strong, no matter what the stock market says, or the clinical trials say, or the bio-Luddites say. Indeed, just like the bio-Luddites, they have turned biotech into myth. For them it is an act of faith that what sounds quacky wacky now won't be quacky wacky tomorrow.

The great uniting belief of the bioutopians is that the future is just about here, that every nano-bio-info-cogno option is coming, and soon. But don't hold your breath. For the moment, we still live, mostly, in houses made of wood and bricks and drive cars with gasoline engines and make electricity with steam. There are, thank goodness, no ray guns. Most people have decided that though it is possible to make their refrigerators talk to the grocery store, they don't necessarily want them to. Thirty years after the start of the War on Cancer, people still get cancer. They die from it by the millions. About 20,000 Americans each year die of influenza and millions of people around the world die of malaria. There is no cure for the common cold. Aging is 100 percent fatal.

Why hasn't the rapture come? Why aren't freaky GM babies popping out of gynoid wombs? Why do people still get old and die of heart disease, and why are kids spending long hours playing video games with titles like "Vice City" instead of using their megabrains to work out the Grand Unified Theory of physics?

As Haseltine said, science is hard. Medicine is even harder. Scientists can do amazing things in labs, but for all that science knows about human biology, there is an awful lot it doesn't know. Just because there are several ways to genetically engineer people does not mean it would work with any regularity. Science has been struggling to add or fix genes in existing people for nearly twenty years and still hasn't managed to move beyond experiments. The future stubbornly refuses to be designed. The biotech experience proves that. So don't look for any grand explosion shot into the bright bio

future. There will be no freaky GM babies, no centenarians running
wind sprints during Big Ten football tryouts. Not any time soon, anyway. They will come, but by the time they do, they will not be
regarded as miracles any more than the polio vaccine. Meanwhile,
you have to have faith.

Haseltine does. He is not a bioutopian. He is a skeptical scientist
and though he believes immortality is possible and desirable, he does
not think it is around the corner. But he has never lost his faith or
his focus. If he makes enemies along the way, so be it. He shrugs off
the resentment many feel toward him, and the snide comments,
both public and private, that people make about his wealth, his hob-
nobbing, his limo. He shed his Vietnam-era liberal guilt over being a
capitalist a long time ago. Being rich and knowing a lot of influential
people, aside from just being more fun, lets him spread the word. "I
am fortunate to be listened to," he said. Meanwhile, the faithful
keep the lamps burning.

"BAD ASS!" China Cat exclaimed.

No shit, seriously, it's awesome, the young man showing off the
teeny-tiny beta test MP3 player assured her. No bigger than a book of
matches and it held, like, practically a whole CD, which was so cool.

"Bad ass!" Her pixie-cut red punk hair shook and the belly ring
through her navel jiggled as she stamped her three-inch-high rubber
platform sandals gleefully on the floor. The teeny-tiny MP3 player
was awesome. China Cat could stuff one of her own L.A. recordings
into that little thing, which was mind-blowing amazing if you
thought about it.

Cat owed her fledgling singing career to technology. Ray
Kurzweil was sponsoring her rise up from L.A. showcase clubs, and
he was doing it with money he made creating and then talking and
writing about technology. Kurzweil had decided to turn himself into
a singer, too, by creating a virtual reality entertainer named Ramona
that he had been using to wow audiences at his speaking engage-
ments for several years. Kurzweil would talk to Ramona, who would
be projected on a big video screen, and then Ramona would crack a
smart-alecky reply and everybody would think how clever it was and
how bad ass technology was getting and ignore the fact that Ramona
seemed a lot like a two-dimensional Andy McCarthy with Kurzweil

playing Edgar Bergen. He had done it again, here at this camp revival meeting disguised as the Extreme Life Extension conference in Newport Beach, California, in late November 2002, and even though the schtick had become hoary everyone was suitably impressed and agreed that pretty much everything Kurzweil said about the coming of the New Age was absolutely true.

Bio-nano-info-cogno. Finally. After all this time the bioutopians had something to say, something concrete with which to evangelize the rest of the world. What a story they could tell! physicist and science fiction writer Greg Benford told them. They were the few, the proud, the enlightened. Who among them, he asked, knew the same Heinlein story he had read as a kid, the one that inspired him? When nearly every hand went up, Benford sighed loudly. "Ah, the Brotherhood." In the past the brotherhood was beleaguered. They talked about death and ways of escaping it, which only reminded people that death was coming and that sure was a big turnoff. Besides, there was already an entire industry devoted to preaching about death and it had gotten into the game long before they had. It was called religion. Religion had a saleable message promising, as it did, everlasting life in paradise. That's why the brotherhood spent years jabbering among themselves. Now though, it was time to escape parochialism because there was a compelling case to be made. "Ours is the first rational solution to death, the nontheological solution." Take this message and spread it, he instructed. Take it to the media—and look at all the media in the back of the room! *Toronto Globe and Mail,* VH1! Swedish TV! Wired News!—attract millions, pass the good word that the movement now had a clear Apostles' Creed: "We believe there is a true destiny up ahead. The techno rapture."

No more depressing rants against the mainstream so typical of past conferences. Science was now their ally. Michael Rose even testified about his own conversion from skeptic to believer in the future immortalization of the human species. Real biologists and influential people supported most of what they had been saying for so long. These people straddled the line between fringe and mainstream, like Rose, Leonard Guarente, Mike West, and of course William Haseltine.

CR dieters, Extropians, nano gurus, New Economy Internet entrepreneurs, science fiction writers. Saul Kent. Cambridge Uni-

versity biologist Aubrey de Grey. Alcor. Bonnie Blood, Extro enthusiast and wife of Ricky "The Dragon" Steamboat, world-famous professional wrestler. They were all here. There was a guy selling cryonics insurance policies, and another one from Morgan Stanley who
could help set up cryonics trusts.

Everyone was excited. Alcor even splurged on party favors for
attendees, slick portfolios imprinted with Alcor's logo, not unusual
for a biotech industry conference, but wildly extravagant for Alcor,
which was so chronically broke the PR materials often looked like
they had been pasted together by a grade school art class. My
favorite Alcor brochure from the early 1990s was fronted by a photo
of a kindly, gray-haired man in a long white lab coat. A woman—a
teacher maybe—and two ten-year-old children, all-American Bobby
and Susie types, were receiving a very interesting tour from the
kindly man, the way kids on a class outing might visit the factory
that makes Fruit Loops, except this kindly older man was gesturing
to the shiny ten-foot-tall stainless steel dewar and explaining, I supposed, the process. "You see, Bobby," I imagined him saying, "first we
cut the heads off . . ." But the cheesy times were over because a couple of wealthy cryonicists had died. One left a million dollars to
Alcor and the other arranged to pay $100,000 every year out of her
estate. Annual membership dues brought in a couple hundred thousand dollars a year to Alcor and gifts only amounted to about a quarter million, so the sudden windfall made an instant impact. Besides,
Alcor couldn't afford to look cheap anymore, not since Ted Williams
and all the attention his arrival had generated. Alcor didn't mind the
hammering by sports columnists decrying what they regarded as the
Splendid Splinter's ignominious storage. For one thing cryonicists
know a lot more about Alan Turing than Richie Allen so they didn't
care what a sports columnist has to say and after thirty-five years of
being called nut job sci-fi geeks, an outraged sports writer is a
buzzing gnat. Any publicity was good publicity and now that Ted
Williams had provided a ton of it, Alcor wanted to look as proficient
and serious as possible.

New Alcor president Jerry Lemler was proud of the revamped
attitude. Lemler came from far outside the old guard, sci-fi enthusiast, cryonicist background. He was a psychiatrist from Nashville by
way of New York City who first became interested in cryonics in

2000 when he walked into a bookstore and picked up Eric Drexler's *Engines of Creation*, one of the foundation documents of nanotechnology. Everything he had learned since left him utterly convinced that with the right application of money and science, human immortality was perfectly achievable, sooner rather than later. With Lemler, amateurism was out, professionalism was in. Science was advancing the state of the art so quickly that soon they'd be able not only to vitrify heads, but whole bodies, turning dead people into six-foot Lladro sculptures. "The old guard set the dream," Lemler said, "but now science must carry the day. Science is now our future." Alcor even had newly professionalized "rescue teams," squadrons of intrepid saw jockeys who would rush to a bedside on a moment's notice so that the prep work could begin at the instant of death. Before, volunteers were handling the procedures and as well intentioned as they were, well, let's just say things did not always go as planned.

The strategy was working. Alcor was beginning to attract younger people weaned on the constant rush of new technology. Twenty-seven-year-old Steven Vachani just signed up. He had made a lot of money during the Internet boom and now commuted between Rio de Janeiro and New York. His buddy, another web entrepreneur who had co-founded a company called PayPal, was a believer in cryonics and sent Vachani to the meeting.

The ultimate goal, of course, was to avoid needing vitrification at all. Alcor wanted to be put out of business. Apparently Lemler did not have long to wait. "If we can all hang in there for another ten years," Kurzweil told the crowd, "we may get to experience the remarkable century ahead." And what a century! Computers will disappear by 2010. By then, "the bridge to the biotech revolution" will have been built and biotech and computer science will merge to create ubiquitous high bandwidth, the reverse engineering of the human brain, nanorobotic implants in our heads, the vast expansion of human intelligence, the exponential extension of the human life span, and then, by 2029, one year earlier than the rapture predicted by FM 2030 . . . The Singularity.

Reaching The Singularity was the prime spiritual quest. "The accelerating growth of evolution never achieves an infinite level, but as it explodes exponentially, it moves rapidly in that direction,"

Kurzweil has written. "So evolution moves inexorably toward our conception of God, albeit never quite reaching this ideal. Thus the freeing of our thinking from the severe limitations of its biological form may be regarded as an essential spiritual quest."

"My generation is the pivotal generation," Kurzweil announced with baby boomer bravado. No more dying the old-fashioned way. Starting now, we could build "a bridge to a bridge to a bridge" and then, starting with a few lucky boomers, biotech would sweep in for the rescue. Kurzweil was doing everything he could to build his own personal bridge. He took 150 supplements per day. He used a special machine to make alkaline drinking water. He ate a special diet designed by an antiaging doctor.

Robert Freitas was just as positive. He talked up the magic of the ultrasmall, as if "respirocytes" were rolling off assembly lines right now. The respirocytes could replicate the entire respiratory system so that you could, if you thought it sounded fun, sit at the bottom of a swimming pool for four hours. Best of all, Freitas announced, he had actually conducted studies of "microbavores," nano germ killing machines. The studies all took place on his computer. There were no such things as microbavores, or respirocytes for that matter, and nobody had any idea how to make them, but Freitas insisted it was all very doable and, as if to prove it, he showed a tricked-out Disney-esque computer graphic program and everyone sat slackjawed as a submicroscopic death star approached a bacteria, released a grappling arm from a silo embedded in its shell, grabbed the bacteria, hauled it into the death star, and then ground the dastardly thing to bits. There was cheering.

Between the new nanotechnology, the advent of artificial chromosomes to eliminate all genetic disease, and the biotech innovation that has already started, human physiology would be rolled back to our late teens, he told the crowd. "Oldness and aging are a disease," he declared, "and you deserve to be cured."

Aubrey de Grey made it sound so simple. All he needed was ten years, a dedicated team, and $100 million and he'd show the world how "negligible senescence" could be engineered into people. De Grey, a theoretical biologist and something of a longevity provocateur who resembles a cross between a young Moses and a member of ZZ Top, insisted that once the ten years were up, and long strides against aging had been made, the world would be so impressed that it would dedicate large portions of the worldwide research establishment to figuring out ways to go all the way to indefinite life extension. As a jump-start, de Grey announced an upcoming summit meeting in England scheduled for September 2003. A long list of top people would be there. Haseltine was a boldface headliner.

The future was sure gonna be a bad ass place to live.

That night I sat with Miller Quarles and his girlfriend Brenda out in the lobby of the Marriott. Miller told me about his career looking at geology all over the world. A guy could have a lot of adventures leading that kind of life and in his eighty-eight years, he had his share. Like the time down in Mexico City when he got into a bridge tournament and his partner was awful pretty, so, naturally, one thing led to another and well, life can be damn good, can't it? Took him years to get through Cal Tech. The Depression was going on and he had to work to pay for his classes from T. H. Morgan and the rest of them, which was why he was so damn proud of that degree and why he wore his Cal Tech belt buckle practically everywhere. Back then they knew that someday we would get control of ourselves. It was just a feeling in the air then, but now, finally, at the very edge of his life, Quarles thought it would happen soon. Sure felt like it could, didn't it? Some pretty amazing science out there. Quarles funded what he could. Not only Geron, but a fella from Geron who had started a telomere-focused company called Sierra Sciences. And he invested a few thousand in another one called GeriGene.

Mike West was still his biggest hope, though. That's why he and Brenda were sitting in the lobby. West was due any minute, flying in from back east and as soon as West walked through those sliding glass doors, Quarles was going to ask for an update on progress. It was nearly ten and West was late. The lobby was quiet and empty. Maybe his plane was delayed. Brenda finally stood up, told Quarles she was going to bed. Miller said he'd sit in the lobby a little while longer. A few minutes later, I told Quarles I was ready to turn in, too. "Okay," he said. "But I think I'll stay up. He ought to be here soon."

He'd wait, he said, as long as it took.

ACKNOWLEDGMENTS

Books just do not get written without help and in the two years or so this book took to create, I've been lucky to have a fair amount of it. These brief notes are certainly not comprehensive.

Writers of such acknowledgments almost always thank their editors, their agents, their spouses. Sometimes such thanks seem pro forma, but be assured that this book would not have existed had it not been for the unstinting faith and prodigious talents of Amanda Cook at Basic Books who made this volume much better than it would otherwise have been, Joe Regal who played Ratso Rizzo to my Midnight Cowboy—even to the point of hustling me through Times Square—and Shelley Alexander who, unlike Amanda and Joe, did not get paid to suffer. Wife-of-writer is a miserable job.

For the past seven or eight years, I have had the good fortune to work with Alex Heard, a smart editor and a terrific writer with a knack for knowing just what turn in the road the culture is about to take. It's no accident that several recent books about biotechnology (among other subjects) contain references to magazine articles Alex originally assigned. This book germinated with such a phone call from Alex, a leap of faith, and several subsequent brave decisions. I'm grateful.

Several staff members of *Wired* magazine have been helpful, too, especially Adam Fisher.

Thanks also to others who have provided assistance, including Hugo Lindgren, Francesca Hayslett, Mitch Polatin, Mike Messerly, Sarah Scully, Cindy Howland, Marianne Leibert, Lynne Elvin, Joyce and Rex Davidson, the library staff of *The New Republic*, the media staffs of *Science* and *Nature*, the public affairs and history personnel of the China Lake Naval Weapons Research Center, the Ripon Historical Society. Curt Pesmen, a brave writer and considerate editor, introduced me to the wonderful world of life extension in the early 1990s. Professors Bart Thurber and Dennis Rohatyn took time to share important insights on philosophy and science fiction. I am indebted to other writers who have displayed generous camaraderie, especially Kristen Philipkowski of Wired News, Sylvia Pagan Westphal of *New Scientist*, Aaron Zitner of the *Los Angeles Times*, and Duff Wilson of the *Seattle Times*. Many of the notes in the back of this book refer to articles written by the science staff of the *New York Times*. That's a tribute.

Jill Herzig threw slow, arching softballs just when I needed an easy hit. As a result, the bills got paid on time.

Nobody has to submit to interviews with a writer, and those who do take some risks. I received remarkable forbearance from nearly all interview subjects, some of whom, like Cynthia Kenyon, Theodore Friedmann, Christopher Heward, Tanja Dominko, Nancy Howar, Dave Kekich, Natasha Vita-More, and Mike West made special efforts to help. Many interviewees are named in this book, many are not, but I am grateful to all for their openness and the sacrifice of their time.

A very special thanks to Rosemarie Hunziker, Ph.D. She not only provided logistical support and encouragement, but proved to be an extremely valuable sounding board, and a critical reader of certain portions of the text, especially those containing biology concepts. All mistakes in the text are mine alone, of course.

Finally, and most important, thanks to William Haseltine. He graciously opened his homes and his life to me, and answered questions with frankness and often with good humor. This book would not have been possible to write without his cooperation.

NOTES

The reporting for this book took place over a period of eight years. These notes do not reflect every reference, interview, or source that influenced its creation.

EPIGRAPHS

ix   "The biologist is the most romantic . . ." "Daedalus, Or Science and the Future," a lecture by J.B.S. Haldane from Haldane's *Daedalus Revisited*, edited by Krishna Dronamraju, Oxford University Press, 1995.

ix   "I sometimes regret . . ." Letter from Salvador Luria to Tracy Sonneborn, February 10, 1944, the Joshua Lederberg Papers, National Library of Medicine.

ix   "In my opinion in the year 2001 . . ." "Memos from D. V." by Diana Vreeland, *The New Yorker*, September 17, 2001.

1: WAITING FOR THE RAPTURE

2   " . . . at around $120,000 . . ." Merkle's address to 1994 A4M meeting; Alcor literature.

6   "Now he and Shaw lived in the Nevada desert . . ." Address by Shaw to the Longevity Boot Camp, April 27–28, 2002.

6   "Haseltine had attended a confab . . ." Haseltine and James Cavanaugh provided the details of the Haseltine–Venter meeting and the run-up to the meeting.

7   "Venter had a reputation for raising hackles within the institutes . . ." *The Gene Wars*, by Robert Cook-Deegan; W. W. Norton and Company, 1995.

7 "Venter and the firm were talking about a lot of money . . ." Interview with James Cavanaugh.

2: THE PROPHET

11 "On February 4, 1923 . . ." I am indebted to the little book, *Haldane's Daedalus Revisited*, by Krishna Dronamraju; Oxford University Press, 1995, for much of the biographical information herein.

13 "in work done by Francis Crick . . ." *The Astonishing Hypothesis: The Scientific Search for the Soul*, by Francis Crick; Simon and Schuster, 1994.

13 "influenced Mary Shelley . . ." *Frankenstein's Footsteps: Science, Genetics and Popular Culture*, by Jon Turney; Yale University Press, 1998, pp. 21–22.

14 "It is possible to believe . . ." Wells's quote was reprinted in *Nature*, February 7, 2002, p. 590.

14 *The Island of Dr. Moreau*, by H. G. Wells; Signet Classics, 1988.

17 *The Golden Bough*, by Sir James George Frazer; Macmillan Publishing Company, 1963.

18 *R.U.R* and *The Insect Play*, by The Brothers Capek; Oxford University Press, 1975.

19 "Joshua Lederberg . . . 'some of the liveliest and most . . .'" Letter from Joshua Lederberg to Helen Spurway, September 11, 1967. (Lederberg Papers, National Library of Medicine).

19 "But *Daedalus* also incited an uproar . . ." Dronamraju.

19 *Brave New World*, by Aldous Huxley; Perennial Classic, 1969.

20 "Galton came to believe . . ." "Can We Learn from Eugenics?" a paper by Daniel Winkler, Ph.D., University of Wisconsin, March 1998; "Statement of the Board of Directors of the American Society of Human Genetics: Eugenics and the Misuse of Genetic Information to Restrict Reproductive Freedom," October 1998; *Controlling Human Heredity*, by Diane B. Paul; Humanity Books, Amherst, N.Y., 1998.

20 *Looking Backward*, by Edward Bellamy; Signet Classic (New American Library edition), 1960.

22 " . . . WISTRAT . . ." History of the Wistar Institute (www.wistar.upenn.edu).

22 "Hammond . . . told an interviewer in 1962" *Animal Production* (1962) volume 4, number 1.

24 "Woodcuts from the early 1700s . . ." Lecture presentation of Robert Winston to International Society of Cytology, May 9, 2002.

24   "During the late 1890s, Thomas Hunt Morgan . . ." Lecture presenta-
     tion of Phil Newmark to Arthur M. Sackler Colloquia of the National
     Academy of Sciences, October 19, 2002.

25   "This work sparked brief excitement when, in 1911 . . ." Turney.

25   "J.B.S. Haldane's wife . . ." Letter from Joshua Lederberg to Hattie
     Gibbs, May 5, 1966 (Lederberg Papers, National Library of Medicine).

3: THE ENDLESS FRONTIER

27   "U.S. Office of Scientific Research and Development . . ." *The Age of
     Science: What Scientists Learned in the Twentieth Century*, by Gerard
     Piel; Basic Books, 2001. And for some information on Vannevar Bush,
     The Billion Dollar Molecule, by Barry Werth; Simon and Schuster, 1994.

26   "William Haseltine was born . . ." Biographical information on Hasel-
     tine and the Haseltine family from interviews with William A. Hasel-
     tine, Florence Haseltine, and "Interview with Florence Haseltine,
     Ph.D., M.D." by the Oral History Project on Women in Medicine,
     Medical College of Pennsylvania, 1977.

30   "the problem of the future . . ." Horowitz to Delbruck (Joshua Lederberg
     Papers—original in the California Institute of Technology Archives).

30   *What Is Life?*, with *Mind and Matter*, and *Autobiographical Sketches*,
     by Erwin Schrodinger; Cambridge University Press, 1967.

31   "Maurice Fox . . ." Interview with Maury and Sally Fox, September 7,
     2002.

32   "Avery wrote to his brother . . ." *The Professor, the Institute and DNA*,
     by Rene Dubos, Rockefeller University Press, 1976.

38   For more on Cold Spring Harbor during the early days of molecular
     biology, see *Genes, Girls and Gamow*, by James D. Watson; Alfred A.
     Knopf, 2001.

40   "'The record of religious beliefs . . ." *The Astonishing Hypothesis: The
     Scientific Search for the Soul*, Crick, p. 258.

40   *The Double Helix*, by James D. Watson; Mentor, New York, 1969.

42   "Gerard Piel . . ." *The Age of Science*, Piel.

44   "RNA Tie Club . . ." *Genes, Girls and Gamow*, by James D. Watson;
     Alfred A. Knopf, 2001.

4: ARISE, LAZARUS LONG!

47   "When Howard Turney was fifty-nine years old . . ." *Grow Young with
     HGH*, by Dr. Ronald Klatz with Carol Kahn; HarperCollins, New
     York, 1997, pp. 64–66.

47 "old friend of Florence Haseltine's . . ." "Effects of Human Growth Hormone in Men over 60 Years Old," New England Journal of Medicine, July 5, 1990, Daniel Rudman et al.

48 "Turney . . . slapped with an injunction . . ." United States Securities and Exchange Commission Litigation Release Number 16425, February 4, 2000.

48 "Dr. Paul Niehans . . ." (www.paulniehans.ch), (www.quackwatch.org).

48 " . . . including a woman named Florence Mahoney . . ." Interview with Deeda Blair.

49 Revolt in 2100 and Methuselah's Children, by Robert A. Heinlein; Baen Publishing Enterprises, Riverdale, N.Y., 1999.

50 Time Enough for Love, by Robert A. Heinlein; G. P. Putnam's (Ace ed.), New York, 1988.

50 "Ettinger had been inspired by . . ." The Prospect of Immortality, 1987 Introduction, by Robert Ettinger. (www.cryonics.org).

51 "By 1950 it was a bestseller." (www.caderbooks.com).

52 "I knew that my youth and health were short lived . . ." "The Failure of the Cryonics Movement," by Saul Kent, Internet posting, April 1998.

53 " . . . the tiny number of people who were seriously interested in it . . ." By 1999, Alcor, the largest cryonics organization, had 460 members with thirty-five in "cryostasis."

53 "FM Esfandiary was . . . captivated . . ." Interview with Natasha Vita-More; "Transhumanism Meets Futurism," by Anders Sandberg, The Transhumanist, November 1999.

54 "Drop acid in the right ways . . ." Your Brain Is God, by Timothy Leary; Renin Publishing, 2001.

54 " . . . Esfandiary, Kent, and Leary were all in California . . ." Interview with Natasha Vita-More.

54 "Leary developed a new slogan . . ." Interview with Gabriel Wisdom.

55 " . . . Nancie Clark picked up an alternative L.A. newspaper . . ." Vita-More.

56 "The smart drugs message appealed to a few members of the San Francisco Bay Area's fledging computer scene." Interview with R. U. Sirius.

56 "Sirius attracted some money . . ." "Mondo 1995," by Jack Boulware, San Francisco Weekly, October 11, 1995.

57 " . . . global brain took on an air of inevitability." "A Globe, Clothing Itself with a Brain," by Jennifer Cobb Kreisberg, Wired magazine, June 1995.

57 "Wired founder Louis Rossetto would be addressing . . ." "Cyberspace

vs. the State," a speech given by Louis Rossetto, February 23, 1996. (www.cato.org).

58   " . . . coalesced in the person of Englishman Max O'Connor." "The Politics of Transhumanism," by James J. Hughes, Ph.D., 2001, Society for the Social Studies of Science.

59   "first cryonics organization in Europe . . ." Interview with Max More and Vita-More.

62   "There was even a brush with the law in 1987 . . ." *Great Mambo Chicken and the Transhuman Condition*, by Ed Regis, Perseus Books, Reading Mass., 1990; *Apocalypse Pretty Soon*, by Alex Heard; Doubleday, N.Y., 1999.

62   "Alcor suspended a television producer . . ." "Dick Jones," by R. Michael Perry, Ph.D. *Cryonics*, 2nd quarter, 1999.

62   "thirty hours elapsed before the suspension process could begin." Vita-More.

5: THE IMMORTAL MR. STEINBERG

68   " . . . a young postdoc named Inder Verma . . ." Interview with Inder Verma.

69   "Theodore Friedmann and Richard Roblin . . ." "Gene Therapy for Human Genetic Disease?" *Science*, March 3, 1972.

69   "Nixon was advised on this policy . . ." Cavanaugh.

70   "On Living in a Biological Revolution," by Donald Fleming, *The Atlantic Monthly*, February 1969.

70   " . . . they called themselves the church . . ." Interview with Jerard Hurwitz.

72   "He wrote of this possibility in 1971 . . ." "Moving Toward Clonal Man," by James D. Watson, *The Atlantic Monthly*, May 1971.

73   "It was featured in *Nature* and Andy became an international celebrity . . ." "Factor Necessary for Ribosomal RNA Synthesis," Travers et al., *Nature*, November 21, 1970. This paper was the culmination of several years' work by Travers and co-workers which was also published in *Nature*. Travers did not respond to repeated attempts to contact him in order to verify Haseltine's account of subsequent events.

74   " . . . competition between himself and other researchers . . ." Interviews with Haseltine and e-mail exchange with Klaus Weber.

74   " . . . the work was turned into a 1972 paper for *Nature* . . ." A short paper was published first: "In Vitro Transcription of *Escherichia coli* Ribosomal RNA Genes," *Nature*, February 11, 1972, followed by the

larger paper, "MSI and MSII Made on Ribosome in Idling Step of Protein Synthesis," *Nature*, August 18, 1972.

76   "Twenty years after events in Cambridge . . ." Interviews with Dennis Kleid, Nancy Hopkins, Inder Verma, et al.

77   "Donald Francis . . . recalled the atmosphere." "The AIDs epidemic in San Francisco: The Medical Response," Oral History Office of the Bancroft Library, 1997 (www.sunsite.berkeley.edu:2020/dynaweb/ teiproj/oh/ science/francis).

78   "The four started a company they christened Cetus." *Making PCR: A Story of Biotechnology*, by Paul Rabinow; University of Chicago Press, 1996. Interview with Paul Berg, Oral History Office of the Bancroft Library, 1997 (www.sunsite.berkeley.edu:2020/dynaweb/ berg). *From Alchemy to IPO: The Business of Biotechnology*, by Cynthia Robbins Roth; Perseus, Cambridge, Mass., 2000.

80   "After the Supreme Court ruled . . ." *Diamond v. Chakrabarty*.

80   "Robert Swanson . . . realized" The oral history office of the University of California's Bancroft Library contains an exceptional interview with Swanson on the founding of Genentech (www.sunsite.berkeley.edu:2020/dynaweb/teiproj/oh/ science/swanson)

80   "Inco and Kleiner-Perkins . . ." International Nickel. Interview with Stuart Feiner, executive vice-president and general counsel.

80   "Kleid had left MIT . . ." Interview with Kleid.

81   "Inco held a series of meetings . . ." Feiner.

81   "It was named after the magic substance . . ." Walter Gilbert isn't so sure the company was named after the substance in *R.U.R.* He had read the play. There was also a lab machine called a biogen that was used to grow cells and the origin of its name may or may not have been inspired by the play. But Gilbert believes the name may also have been simply a coincidence—a desire of the investors to unite the "gen" of Genentech with "bio."

82   "Robert Foman, the president of E. F. Hutton . . ." Interview with Al Scheid; interview with John Baxter.

83   " . . . companies got started based on scientists walking out of labs with test tubes in their pockets . . ." The most famous case involved Genentech itself, when a scientist claimed he took genetic material used to make HGH from the labs of the University of California at San Francisco to help produce Genentech's HGH. As reported in *The Scientist*, (June 10, 2002) biologist Peter Seeburg testified in court that "that's the way we did it twenty years ago." Genentech disputed Seeburg's account, but, without admitting wrongdoing, agreed to

settle and intellectual property dispute related to this alleged incident by paying UCSF $200 million.

83 " . . . a twenty-four-year-old aspiring songwriter in New York named David Blech . . ." "The Financier," part of the series "Uninformed Consent" by Duff Wilson and David Heath, *Seattle Times*, March 11–15, 2001.

83 "Goodman worked at a venture firm . . ." Interview with Ed Goodman.

84 "Investing in what was now called Cambridge Bioscience was a no-brainer for D. Blech and Company." Interview with David Blech.

84 "Gallo recalled in a *Science* memoir . . ." "The Early Years of HIV/AIDs," by Robert Gallo; *Science*, November 29, 2002.

85 "In April 1984, Cambridge Bioscience announced . . ." "Cambridge BioScience Seeks to Patent Tests on Likely AIDs Virus," by Bob Davis, *Wall Street Journal*, April 26, 1984.

85 " . . . it was granted the exclusive license . . . to gp120" "Entrepreneur Briefs," *The Scientist*, May 30, 1988; interview with David Blech.

85 "Haseltine blamed the CEO . . ." Interview with Haseltine. His view was shared, somewhat, by David Blech.

86 "finally came crashing down in 1994 amid a scandal . . ." "Ex Insider at Cambridge Biotech Tells How Firm Allegedly Used Bogus Sales," by David Stipp, *Wall Street Journal*, July 26, 1994.

87 "I have this theory that death is a genetic disease . . ." "Wallace Steinberg; "Laying Pipe for the Fountain of Youth," by Gina Kolata, *New York Times*, November 1, 1992.

87 "Wally took me aside . . ." Interview with Michael West.

87 "He totally believed in himself . . ." Interview with Billy Walters.

89 "Steinberg would perform a shtick . . ." Cavanaugh.

89 "own American science . . ." Anonymous source.

90 "He could be dark and cruel . . ." Interview with Nancy Howar.

90 "There were no limits . . ." Interview with Mona Geller.

90 "Scientists who later met Steinberg . . ." Interviews with Denise Faustman, Michael Zasloff, Florence Haseltine, et al.

90 "I would say he was definitely a competitor . . ." Interview with Jay Short.

91 "Deeda Blair was born . . ." Biographical information, interview with Deeda Blair.

92 "In 1942, Mary Lasker . . ." The Lasker Foundation (www.laskerfoundation.org).

94 "'He looked at me and said . . ." Gilbert does not recall this conversation or this particular meeting with Deeda Blair.

94 "Deeda could walk through the NIH . . ." Anonymous source.

94 " . . . Nancy Howar . . . suggested they form a two-person art consulting business." Howar.

96 "scientists were lured . . ." Faustman, Zasloff.

97 "Deeda ran interference, assuring skeptical administrators . . ." Faustman (Faustman's company, Diacrin, was one of those formed out of the Healthcare Ventures–Harvard deal); "Wallace Steinberg . . ." Kolata.

97 " . . . not one of Steinberg's companies has been successful." One company, MedImmune, came close in 2002 with total revenue of over $800 million. Again, the $1 billion target was Steinberg's as reported by Kolata. Some companies certainly have made money for investors through a variety of mechanisms like stock run-ups or buyouts by other firms.

98 "Watson did not want to use Venter's technology . . ." *Gene Wars*, Cook-Degan; interview with Haseltine.

99 "Haseltine was not Steinberg's only option." Steinberg also interviewed biotech executive Tom Wiggans. Interview with Wiggans.

100 "Enormous power has been inherited . . ." *The Recombinant DNA Controversy: A Memoir*, by Donald S. Fredrickson, M.D., American Society for Microbiology Press, Washington, D.C., 2001.

101 "There were to be three sister companies feeding off TIGR . . ." Haseltine. Short.

101 " . . . it sent George Poste to Florida . . ." Cavanaugh.

102 "According to an unofficial version . . ." Interview with Alan Walton.

6: WAY OUT WEST

103 " . . . His boss had to twist Harley's arm." West.

104 "West had been on the case for fifteen years or so . . ." Biographical information, interviews with West.

107 "Alex Comfort . . . had written a little . . ." "The Prevention of Ageing in Cells," *Lancet*, December 17, 1966; "Feasibility in Age Research," *Nature*, January 27, 1968 (among other publications).

108 " . . . a colleague of Harley's named Carol Greider . . ." "Identification of a Specific Telomere Terminal Transferase Activity in Tetrahymena Extracts," Greider et al., *Cell*, December 1985.

109 " . . . among the other people who tried to promote cryonics . . ." "The Failure of the Cryonics Movement," Kent.

109 "Quarles . . ." Biographical information, interview with Miller Quarles. Interview with West.

111 "Tension between what West . . ." Interview with Tom Okarma, CEO of Geron. Interview with West.

112 "The first surrogates for these cells came from odd cancers . . ." Biologist Leroy Stevens is generally credited with the first work with teratomas.

112 " . . . sad attempts at building real systems." Interview with Martin Evans.

114 " . . . they wanted to create the real McCoy . . ." Interview with Peter Andrews.

115 " . . . the developmentalists had no real interest in or expectation . . ." Interview with John Gurdon.

116 " . . . they did come from a place where nobody was looking . . ." This is a widely held view among ag reproductive specialists. Interviews with Neal First, Jose Cibelli, Tanja Dominko, Ian Wilmut, et al.

116 " . . . Wilmut's . . . lackluster college career . . ." Interview with Ian Wilmut.

117 " . . . a geep as it turned out." This work done by Steen Willadsen.

118 "But ES cells proved elusive . . ." Wilmut.

118 "Campbell, intrigued by cloning . . ." *The Second Creation: Dolly and the Age of Biological Control*, by Ian Wilmut, Keith Campbell, and Colin Tudge; Farrar, Straus and Giroux, New York, 2000.

118 "babies are born young." Interview with Calvin Harley.

119 " . . . a Singaporean named Arrif Bongso announced . . ." "Isolation and Culture of Inner Cell Mass Cells from Human Blastocysts," Bongso et al., *Human Reproduction*, November 1994.

119 "Gearhart was getting death threats for his trouble . . ." "The Troubled Hunt for the Ultimate Cell," by Antonio Regalado; *Technology Review*, July–August 1998.

120 "It was really fun . . ." Interview with Phillipe Collas.

122 "So he sent some to Peter Andrews . . ." Andrews.

123 " . . . Haseltine said he was discussing a collaboration with Geron." "Learning the Cells' Language," by Nicholas Wade, New York Times News Service, December 16, 1998.

7: Bring On the Inquisition

126 "One man who pled for a go-slow approach . . ." *The Recombinant DNA Controversy: A Memoir* by Donald S. Fredrickson; ASM Press 2001.

127 *A Matter of Life*, by Robert Edwards and Patrick Steptoe (now published by *Reproductive Biomedicine Online*, 2001).

127   " . . . seen as something of an anachronistic zealot . . ." The disdain
       was reciprocated. According to Kass ally Francis Fukuyama, bioethi-
       cists in the mainstream had been "hopelessly co-opted by the biotech-
       nology industry."

128   *Island*, by Aldous Huxley; Harper and Row (Perennial Classics ed.),
       New York, 1972.

128   "These are the last people . . ." Interview with George Annas.

129   *Life, Liberty and the Defense of Dignity*, by Leon Kass; Encounter
       Books, San Francisco, 2002.

129   "Robert Welch's John Birch Society Blue Book . . ." *The Blue Book of
       the John Birch Society*, by Robert Welch, (fourth printing) 1961. For
       example: "we have the equally important longer range job of ending
       this mass psychological flight towards amorality; and of restoring con-
       vincing reasons for men once again to strive to live up to moral and
       humanitarian ideals. Otherwise, there is no chance of saving our
       Christian-style civilization from self-destruction . . . Already in hun-
       dreds of other ways, the Communists are rubbing out or making more
       and more shadowy the lines of disagreement and the once sharp dif-
       ferences between our ways and theirs. The movement is smooth,
       widespread, continuous, insidious, and powerful."

129   *Our Post Human Future: Consequences of the Biotechnology Revolution*,
       by Francis Fukuyama; Farrar, Strauss and Giroux, New York, 2002.

130   "Absolutely . . ." Interview with Francis Fukuyama.

130   "we're not ready . . ." Annas interview.

130   "Magazine editor Tina Brown . . ." "As Talk Magazine's Run Ends, the
       Spin Begins," by David Carr, *New York Times*, January 20, 2002.

130   " . . . maybe [male] DNA says . . ." "They Conquered, They Left," by
       Alex Kuczynski, *New York Times*, March 24, 2002.

130   "Martin . . . might feel guilty . . ." "Charlie Sheen's Redemption Helps
       a Studio," *New York Times*, February 4, 2002.

130   "ex-substance abuser Martin . . ." "The Sheen Team" (Martin and Char-
       lie Sheen Interview) by Mary Murphy, *TV Guide*, February 25, 2002.

130   " . . . the clown's rogue gene." "Bring Me Sunshine," by John Lahr;
       The New Yorker, January 21, 2002.

130   "Is there a special gene among intellectuals . . . ?" "Why Are Deep
       Thinkers Shallow About Tyranny?" by Eric Alterman; *New York Times*,
       November 10, 2001.

130   " . . . the shopping gene . . ." "A Publication Date," by Paul Rudnik
       and Alex Witchel, *Fashions of the Times*, Fall 2002.

130   " . . . Books now had DNA . . ." "the book's DNA seems to be made up

solely of cigarette butts . . ." "Raw Bar," by Dwight Garner (a review of *Crawling at Night, New York Times Book Review*, April 29, 2001).

130 "Houses had DNA." "Even a distinguished modern house can make room for the next generation—especially when the design DNA is there." Pilar Viladas quoted in the *New York Times Sunday Magazine*.

130 " . . . walking armature of DNA." "The Guardians," by John Updike; *The New Yorker*, March 26, 2001.

131 "Bradley Rubenstein shares . . ." "DNART," by Peter Schjeldahl, *The New Yorker*, October 2, 2000.

131 "Three Tales . . ." "Heartburn," by John Lahr, *The New Yorker*, January 6, 2003.

131 "The bio-Luddites should exploit cloning fear . . ." "Why We Should Ban Human Cloning Now. Preventing a Brave New World," by Leon Kass, *The New Republic*, May 17, 2001.

132 "as New York's Catholic cardinal . . . famously declared . . ." "(You)2" by Brian Alexander, *Wired* magazine, February 2001.

133 "Harry Griffin . . ." Interview with Harry Griffin.

134 "hoo-ha . . ." Wilmut interview.

135 " . . . cloners themselves would not particularly care if people were cloned . . ." Interviews with Dominko, Willadsen, Alan Trounson, Michael Bishop. Wilmut has somewhat different views on the subject, best described as "uneasiness," or "misgivings."

135 " . . . if the first clone is born . . ." "DNA Pioneer Backs Cloning of Humans," by Severin Carrell, *The Independent*, April 13, 2003.

137 " . . . CuresNow . . ." (www.curesnow.org)

137 "Which one of my children would you kill?" "Stem Cell Debate in House Has Two Faces, Both Young," by Sheryl Gay Stolberg, *New York Times*, July 18, 2001.

139 "What really ought to give us pause . . ." Charles Krauthammer, *Time*, February 5, 2001.

139 "I think it is playing God . . ." Interview with Cliff Stearns.

141 "an industry of death . . ." G.O.P. Leaders in House Fight Stem-Cell Aid," by Robert Pear, *New York Times*, July 3, 2001.

141 "I have learned the hard way . . ." Interview with Douglas Melton.

144 " . . . count them on one hand, two at the most . . ." Melton. Also, "New Stem Cell Issue," by Sheryl Gay Stolberg, *New York Times*, September 3, 2001, and interviews with Tom Okarma.

144 "The American Red Cross . . ." "Red Cross Shifts, Rejects Pioneering Stem Cell Grant," by Aaron Zitner, *Los Angeles Times*, February 8, 2002.

146 "Foster . . . laughed . . ." Interview with Roland Foster.

147 " . . . a class of superhumans . . ." "Bush's Advisers on Ethics Discuss Human Cloning," by Sheryl Gay Stolberg, *New York Times*, January 18, 2002.

147 " . . . editorialized *Nature* . . ." "Morality, Prejudice and Cloning," *Nature*, January 24, 2002.

147 "Four commission members . . ." "Harmful Moratorium on Stem Cell Research," Rowley et al., *Science*, September 20, 2002.

149 "could be damn cruel . . ." *Engineering the Human Germline: An Exploration of the Science and Ethics of Altering the Genes We Pass to Our Children*, edited by Gregory Stock and John Campbell, Oxford University Press, New York, 2000.

149 "We are the last five minutes of evolution . . ." Robert Edwards, lecture to the American Society of Reproductive Medicine, 2001, Orlando, Fla.

149 " . . . in 1989 two English scientists . . ." "Biopsy of Human Preimplantation Embryos and Sexing by DNA Amplification," Handyside, Winston, et al., *Lancet,* February 18, 1989.

150 "Doctors saw . . . PGD" Interviews with Mark Hughes, Yuri Verlinsky, Santiago Munne.

150 "2002 journal paper by a leading PGD expert . . ." "Preimplantation Genetic Diagnosis for Cancer Predisposition," by Svetlana Rechitsky et al., *Reproductive Biomedicine Online*, September–October 2002.

151 " . . . we'll stop making the popular sex . . ." Interview with Robert Edwards.

152 "Simpson wanted to break free . . ." Interview with Joe Leigh Simpson.

152 *Redesigning Humans: Our Inevitable Genetic Future*, by Gregory Stock; Houghton Mifflin Company, New York, 2002.

152 " . . . began agitating for its consideration . . ." Interview with Gregory Stock.

152 "Most ethical disagreements have been vaporized . . ." "Public Objections to Designer Babies and Cloning in USA: Not Quite What Was Expected," by Joe Leigh Simpson and Robert Edwards, *Reproductive Biomedicine Online*, March 2003.

153 "Nancye Buelow was forty-nine . . ." Interview with Nancye Buelow.

154 "During a meeting of Bush's bioethics advisory council . . ." (www.bioethics.gov/transcripts/jan03/session3.html).

156 "transhumanists . . . next great threat to human dignity . . ." "The Transhumanists: The Next Great Threat to Human Dignity," by Wesley J. Smith, *National Review* September 20, 2002.

156   *The Name of the Rose*, by Umberto Ecco; Harcourt Brace Jovanovich, San Diego, 1983.

8: WATER INTO WINE

160   "an estimated $82 million as her share . . ." "Famous Splits," by Mike Hoffman, *Inc.* magazine, September 1, 1998; "Two Halves of a Whole," by Nina Moukheiber, *Forbes*, November 1992.

163   " . . . a paper by Mike West's Advanced Cell Technology . . ." "Somatic Cell Nuclear Transfer in Humans: Pronuclear and Early Embryonic Development," by Jose Cibelli et al., *eBiomed, The Journal of Regenerative Medicine*, November 26, 2001.

164   "Members of *e-Biomed*'s advisory board announced . . ." John Gearhart, Davor Solter, and Robin Lovell-Badge. "Cloning Claims Challenged," by Jeffrey Krasner, *Boston Globe*, January 30, 2002.

164   "Wilmut, among others . . ." Wilmut interview.

165   " . . . Publish and Be Damned . . ." *Nature Biotechnology*, January 2002.

165   " . . . In an eloquent and impassioned letter . . ." "Transparency in Public Relations," Robert Lanza et al., Nature Biotechnology, February 2002.

166   "I asked that *US News* not learn about what, if anything, we had done with human nuclear transfer . . ." *US News* reporter Joannie Fischer disputes this. According to Fischer, she was repeatedly asked by ACT's leadership to visit the company specifically for the purpose of reporting on the cloning experiments.

167   " . . . a man who handed out business cards . . ." Interview with Rael.

168   "disgorged Michael Rennie in *The Day the Earth Stood Still* . . ." Visit to UFOLand (Raelian headquarters), Valcourt, Quebec.

168   "Claude and the spaceman got along well . . ." Raelian literature.

169   "to dislike the cold weather . . ." Rael interview.

170   "We are so eager . . ." Interview with Buissellier.

170   "Mark and Tracy Hunt . . . Nitro, West Virginia . . ." "Cloned in the USA: Attempt to Clone Human Being in Secret West Virginia Lab Revealed," by Joe Lauria, *London Times*, August 12, 2001.

170   " . . . would have been a miracle . . ." Lauria; "Cloning Presents a Bold Frontier to These Guys, Too," by Dennis Roddy, *Pittsburgh Post-Gazette*, August 11, 2001.

170   " . . . there are some cells . . ." Boissellier interview.

171   "Wicker had joined the New York chapter . . ." Biographical information provided by Randy Wicker and documents from his archives.

172   " . . . columnist Jack O'Brian . . ." "Jack O'Brian Says," *New York Journal American,* July 9, 1962.

174   "Eibert saw cloning . . ." Interview with Mark Eibert.

174   "The other board members included . . ." Summary of the board of directors' meeting of the Human Cloning Foundation, February 11, 2000.

175   " . . . he was actually James Byrne . . ." Meeting with Byrne.

176   " . . . One scientist who communicated regularly with Wicker . . ." His story was obtained through interviews with him, his girlfriend, Wicker, and various communications over a period of two years.

178   "I personally believe that in time . . ." Gurdon interview.

179   "Cloning is a magic trick . . ." Dominko interview.

180   "Experiments done in the 1980s by Helen Blau . . ." "Cytoplasmic Activation of Human Nuclear Genes in Stable Heterocaryons," Helen Blau et al., *Cell,* April 1983.

181   "One company was trying . . ." Intercytex.

182   "We are embarking on a great enterprise . . ." Reprinted in *eBiomed* as "The Emergence of Regenerative Medicine: A New Field and a New Society," by William A. Haseltine, June 7, 2001.

182   "As proof of what his own company could do . . ." Presentation at the Society for Regenerative Medicine meeting, December 2, 2001.

183   "No rock hard boob syndrome . . ." "Don't Die, Stay Pretty," by Brian Alexander, *Wired* magazine, January 2000.

184   " . . . had identified antiaging as an emerging space . . ." Interview with Charles Arday of D. E. Shaw.

186   "science fact and science fiction . . ." Interview with Gordon Lithgow.

188   "Melov, an avid sci-fi enthusiast . . ." "A Lease on Life," by James Meek, *The Guardian,* June 30, 2001.

188   "When the paper announcing the work . . ." "Extension of Life-span with Superoxide Dismutase/Catalase Mimetics," by Simon Melov et al., *Science,* September 1, 2000.

189   "All I am saying is . . ." Interviews with Cynthia Kenyon.

190   "This life span extension . . . Kenyon wrote in Nature . . ." "A *C. elegans.* Mutant that Lives Twice as Long as Wild Type," by Cynthia Kenyon et al., *Nature,* December 2, 1993.

192   "Bayley was an admirer of Wally Steinberg . . ." Interview with Cindy Bayley.

193   "Elixir hoped its drugs would be approved by a similar route . . ." Interview with Ed Cannon.

194 "Death is an outrage!" Freitas presentation to Alcor Extreme Life Extension Conference, 2002.

196 "When they're not busy cloning . . ." "Skintight Genes," by Patricia Reynoso, *W*, November 2002.

196 "Donna, for example . . ." Donna's story was obtained over a period of four years.

198 "Primo 3M+" (www.natasha.cc/primo.htm)

9: POP! GOES THE RAPTURE

201 "Andrews falls into this last category . . ." Interviews and correspondence with Peter Andrews.

205 " . . . even predicting a backlash . . ." Interview with Gail Martin.

205 "In a not-too-surprising announcement . . ." "Notebook," *The Scientist*, July 7, 1997.

206 "Craig Venter, head of TIGR, says HGS's threat . . ." "HGS-TIGR Splits, Opportunity Knocks," *Nature Biotechnology*, August 1997.

207 " . . . be the first of many . . ." Transcript of The Motley Fool Radio Show, March 6, 2000.

209 "He's the only one who thinks so!" Press conference at the American Society for Gene Therapy meeting, 2001.

209 "almost all of them were higher than the 30,000 or so . . ." For example, a Cold Spring Harbor researcher named Michael Zhang estimated that there may be something on the order of 56,000.

209 "Some people just like to argue . . ." Interview with Craig Rosen.

211 "But she also said he had a 'difficult' personality . . ." Oral History Project on Women in Medicine, Medical College of Pennsylvania, 1977.

211 "One of the questions I ask myself . . ." Anonymous source.

212 "Bill is respected for his intellect . . ." Ed Goodman interview.

212 "I think they lost the bet . . ." Jerry Cacciotti of Strategic Decisions Group quoted in "HGS Battles the Pipeline Blues," by Malorye Branca, *Bio-IT World*, June 12, 2002.

212 " . . . a handful, made a profit . . ." "The Changing Norms of the Life Sciences," by Peter Shorett, Paul Rabinow, and Paul R. Billings, *Nature Biotechnology*, February 2003: "Besides the five largest players—Amgen . . . Biogen . . . Chiron . . . and Genzyme—which account for over one-third of industry revenues, most of the 2,000 biotech companies worldwide have yet to generate a substantial revenue generating product."

213 " . . . reported the investment ranking company, Morningstar . . ."

"Splicing the Biotechnology Industry," by Jill Kiersky; Morningstar.com, October 9, 2002.

214 "Some engineered tissues, like ureters . . ." Presentation by Anthony Atala at Second Annual Society for Regenerative Medicine meeting.

214 "The penises were for rabbits . . ." "Tissue Engineers Grow Penis in the Lab," by Sylvia Pagan Westphal, *New Scientist*, September 11, 2002.

216 "By February 2002, the government . . ." The story of ACT's dealings with the government came from interviews with West, Dominko, an anonymous source close to one of the government agencies involved, and news reports.

217 "Pitts grandstanded . . ." Press release dated May 14, 2002, issued by Pitts's office.

217 " . . . Pitts told the *New York Times* . . ." "Cloning Company Misspent Federal Money, Auditors Say," by Sheryl Gay Stolberg, *New York Times*, May 14, 2002.

217 "At least one other antiaging organization West approached . . ." Exeter Life Sciences.

10: The Rapture Rides in a Limo

224 "As Haseltine wrote of bio-Luddite Francis Fukuyama's book . . ." "Less Than Human: First Francis Fukuyama Ended History, Now He Wants to Stop Science Too," by William A. Haseltine, *The American Spectator*, March–April 2002.

224 "People say it would be terrible if we made all girls pretty . . ." "Stupidity Should Be Cured, says DNA Discoverer," New Scientist, February 28, 2003 (reporting on a Channel 4 documentary).

225 "From across the table . . ." During Longevity Boot Camp, April 27–28, Las Vegas, Nev.

230 "Hey, everybody wants to look good . . ." "The Man Who Makes Pictures Perfect," by Kate Betts, *New York Times*, February 2, 2003.

231 "Ellison was sold . . ." "Business Week 50: Masters of Innovation: Bioinformatics," by Ellen Licking, John Carey, and Jim Kerstetter, *BusinessWeek*, April 7, 2001.

231 "Kekich suffered a fluke spinal injury . . ." Interviews with David Kekich.

232 "but they were dubious . . ." Kekich presentation to AGE Conference, June 2002.

232 " . . . the foundation had amassed net assets . . ." Financial information for the Life Extension Foundation and Saul Kent's salary from Internal Revenue Service Form 990—Year 2000.

232 " . . . people frozen before the late 1990s were goners . . ." Interview with Alcor president Jerry Lemler.

233 "nothing left but your big toe . . ." Alcor Extreme Life Extension Conference, 2002.

233 "Its two leaders even gained . . ." Greg Fahy and Brian Wowk, whose research papers, especially on organ preservation, have appeared in publications like *Cryobiology* and *Transplantation*.

233 "Kronos . . ." Tour provided by Christopher Heward.

234 "This was the 'Kronos Way' . . ." Interviews with Heward, Jonathan Thatcher.

236 "Sperling began life . . ." Interview with Sperling; *Rebel with a Cause*, by John Sperling; John Wiley & Sons, New York, 2000.

238 " . . . published a paper in *Science* about daf-2 worms . . ." "Daf-2, an Insulin Receptor-like Gene That Regulates Longevity and Diapause in *Caenorhabditis elegans*," Kimura et al., *Science*, August 15, 1997.

239 " . . . hired a young healthcare executive . . ." Jonathan Thatcher.

239 "Coles plugged Exeter . . ." Heward interview.

240 "When the NIA was first formed, the other institutes laughed . . ." Interview with Cutler.

241 "Haseltine was right . . ." Interview with Harman.

243 "There was no normal, not in the religion of biotechnology . . ." This is a widely held view, not only among bioutopians. One of the most intriguing debates in bioethics has been how to distinguish between therapy and enhancement. Theodore Friedmann, for one, argues that the distinction has become meaningless.

243 " . . . American Academy for the Advancement of Science . . . issued a report . . ." "Human Inheritable Genetic Modifications," September 2000.

244 "Converging Technologies to Improve Human Performance." The report was issued June 2002.

245 "zinc fingers . . ." Commercialized by a company called Sangamo Bioscience.

246 "DARPA . . . had an ongoing program . . ." (www.darpa.mil).

246 " . . . could exist in near stasis . . ." Interview with a DARPA official.

246 "Scientists at Merck created mice with better memories . . ." Presentation at BIO 2001 by Gerry Dawson.

246 "Scientists at the Scripps Research Institute . . . created a whole new species . . ." "Generation of a Bacterium with a 21 Amino Acid Genetic Code," Mehl et al., *Journal of the American Chemical Society*, January 29, 2003.

247 " . . . the World Anti-Doping Agency . . . was pretty sure . . ." Theodore Friedmann interview. Presentation by Johan Olav Koss, University of California at San Diego, 2001.

248 " . . . freakish fusing of two genes . . . twenty-five million years ago." "The Tre2 (USP6) Oncogene is a Hominid-specific Gene," by Charles A. Paulding et al., *Proceedings of the National Academy of Sciences*, March 2003.

248 "Mrs. J'Noel Gardiner . . ." "Suit Over Estate Claims a Widow Is Not a Woman," by Jodi Wilgoren, *New York Times*, January 13, 2002.

249 "Make no mistake about it . . ." E. O. Wilson interview in the *Wall Street Journal*, January 1, 2000.

251 "We need to do an unlikely thing . . ." *Enough: Staying Human in an Engineered Age*, by Bill McKibben; Henry Holt and Company, New York, 2003.

251 "Every generation has a lower threshold . . ." Interview with Ethan Bier.

251 "The earth belongs in usufruct . . ." Letter from Thomas Jefferson to James Madison, September 6, 1789. "I say the earth belongs to each of these generations, during its course, fully, and in their own right."

252 " . . . about 20,000 Americans die of influenza . . ." Centers for Disease Control and Prevention.

253 "Bad ass!" Alcor Extreme Life Extension Conference, 2002.

255 " . . . a couple of wealthy cryonicists had died . . ." Lemler interview.

255 "Annual membership dues . . ." Internal Revenue Service Form 990, Year 2000.

257 "Kurzweil has written . . ." (www.kurzweilai.net/articles/art0358.html).

INDEX

Advanced Cell Technology (ACT),
124, 139, 163–66, 215, 229,
242; disintegration of, 215–18
aging: free radical theory of, 239;
progeria as a surrogate model
for, 107; as a progressive
disease, 107
AIDS research, 84–86
Alafi, Moshe, 78–79
Albert and Mary Lasker
Foundation, 92
Alberts, Bruce, 140
Alcor Life Extension Foundation, 2,
59, 62, 232, 255–56
Allen, Woody, 53
Alzheimer's Foundation, 148
Amazing Stories, 25
American Academy for the
Advancement of Science,
243–44
American Academy of Anti-Aging
Medicine (A4M), 1–2
Amgen, 220
Anderson, W. French, 96, 97, 213
Andrews, Peter, 114, 115–16, 122,
201–5, 214
Androderm, 235
Annas, George, 128
Antinori, Severino, 167

Are You Transhuman? (FM 2030),
57
Aristotle, 24
artificial insemination, 22–24
Asilomar, 126, 146
Atala, Anthony, 163
Atlas Shrugged (Rand), 59–60
Avery, Oswald, 32–33; paper on the
discovery of DNA, 33

Bacon, Francis, 12, 25
Bacon, Roger, 12
bacteriophages, 32, 38–39, 195
Baltimore, David, 68, 75
Bautz, Ekkehard, 73
Baxter, John, 82
Bayley, Cindy, 192
Beacon (More), 131
Beadle, George, 29–30
Beckwith, Jonathon, 68
Bellamy, Edward, 20–21
Benford, Greg, 254
Bensalem, 12–13
Berg, Paul, 75, 79, 126, 145; paper
on recombinant DNA, 79
Biogen, 81, 89, 94, 220
bio-Luddism, 126, 127–29, 195,
229, 230, 247, 251–52; and
"Brave New World," 153;

bio-Luddism (*continued*)
    enshrinement in the federal
    government, 143–44;
    exploitation of cloning fear,
    131, 136–37, 139–41, 166–67;
    failure to explain their position
    clearly, 154–57; opposition to
    life extension, 193; and the
    "slippery slope," 153
Biometrics, 240
biophysics, 67–68
biotech industry, 64, 77–78;
    beginnings of, 183–84; fall of,
    212–13; heyday of, 207
bioutopianism, 153–54, 252
"Birthmark, The" (Hawthorne), 147
Bishop, J. Michael, 145
Blair, Deeda, 91–99, 213, 219;
    board memberships of, 92–93;
    friendship with Mary Lasker,
    92–93; marriage to William
    McCormick Blair Jr., 92; and
    Wallace Steinberg, 94–97; and
    Wesley Associates, 94–97; and
    William A. Haseltine, 95–96
Blair, Tony, 141, 142
Blair, William McCormick, Jr., 92
Blau, Helen, 180
Blavatsky, Madame, 52
Blech, David, 83, 85, 86
"Blighters" (Sassoon), 15–16
Boisellier, Brigette, 169–71
Bongso, Arrif, 119
Borden, John, 137
Botstein, David, 68
Boyer, Herbert, 79
Brand, Stewart, 57
Brave New World (A. Huxley), 19,
    124, 127, 144
Brenner, Sydney, 67–68, 184–85,
    189
Brown, David L., 109
Brown, Louise, 132
Brownback, Sam, 145
Buck, Beryl, 185
Buck, Leonard, 185

Buck Institute for Aging Research,
    185–86
Buelow, Nancye, 153
Bush, George W., 137, 141,
    143–44, 147, 163, 216
Bush, Vannevar, 27
*Butterfly Landscape, The Great
    Masturbator in Surrealist
    Landscape with DNA* (Dali),
    131
Byrne, James, 168, 175

C. elegans, 184–92, 238
California Biotech, 82
calorie restriction (CR), 4, 107, 238
Calvin, Melvin, 44
Cambridge Bioscience, 84–86;
    name change to Cambridge
    Biotech, 86; and Recombigen,
    85
Campbell, Keith, 118, 121
Cannon, Ed, 193–94
Cape, Ronald, 78
Capecchi, Mario, 115
Capek, Josef, 18
Capek, Karel, 18
Card, Andrew, 143
Cavanaugh, James, 69, 87, 88
Cc (Copy cat or Carbon copy), 177,
    235, 242
Celera, 206, 208, 213
cell lines: and agricultural
    scientists, 23, 116–18; EC
    cells, 113–16; ES cells (*see*
    stem cells); HeLa, 108
cellular therapy, 48
Center for Developmental Biology,
    147
Cetus, 78–79, 80
Chamberlain, Owen, 44
Cheney, Dick, 143
China Lake Naval Weapons Center,
    33–37
Cibelli, Jose, 165–66, 215–16, 217
Clark, Nancie. *See* Vita-More,
    Natasha

Clinique La Prairie, 48
Clonaid, 169, 170
Clonapet, 169
Clone Rights United Front
    (CRUF), 174
cloning, 2, 118, 120, 128, 131–33,
    134–37, 140, 179, 215; in
    China, 147; in India, 148; in
    Japan, 147; lack of idea how to
    make it useful to people,
    213–15; in Singapore, 148;
    therapeutic cloning, 137, 145
CLP, 103
co-enzyme Q10, 3
Cohen, Eric, 129, 164
Cohen, Stanley, 79
Cold Spring Harbor, 21, 38–39,
    195, 220
Coles, Stephen, 3, 226, 227,
    228–29, 229–30, 239
Collins, Francis, 208, 209
Comfort, Alex, 107
"Converging Technologies to
    Improve Human Performance"
    workshop, 244–45
"Creation of 'Artificial Life,' The," 25
Crick, Francis, 7, 8, 31, 40–41, 44,
    67, 70; paper on DNA, 41;
    view of religion, 40; view of the
    soul, 13
cryonics, 1, 2, 4, 52–53, 62;
    crudeness of twentieth-century
    methods, 232–33; founding
    document of, 52; and "neuros,"
    2; and vitrification, 233
Cryo-Span, 53
Cure Old Age Disease Society
    (COADS), 109–10
CuresNow, 137
Cutler, Richard, 4, 239, 240–41,
    250
Cyagra, 215

D. Blech and Company, 83, 84, 86
"Daedalus or Science and the
    Future" (Haldane), 11–12,
    18–19; publication in booklet
    form, 19
Dali, Salvador, 131
Darwin, Charles, 12, 16, 20, 21
Dawkins, Richard, 60–61
D. E. Shaw & Co., 184
de Chardin, Pierre Teilhard, 56–57,
    245
de Grey, Aubrey, 255, 257–58
de Vries, Hugo, 21
DeCode Genetics, 192
Defense Advanced Research
    Projects Agency (DARPA),
    245–46
Delbruck, Max, 30, 70; and "The
    Phage Course," 38–39
Descartes, René, 13
developmental biology, 24, 48, 112,
    113–15
dietary supplements, 5–6, 55–56,
    62–63
Diversa, 90, 101, 206
DNA, 7, 32–33; cDNA, 68; DNA
    technology, 63; and fraying of
    telomeres, 108; and
    hybridization, 44; as an icon,
    130–31; recombinant DNA,
    79; structure of, 39–41, 67.
    See also RNA
Dolly, 121–22, 124, 133–34, 178;
    death of, 215; as sign of the
    Second Coming for the
    Raelians, 168, 169, 170
Dominko, Tanja, 179–81, 195, 216,
    217, 246
Double Helix, The (Watson), 40
Draper, John, 201–2
Drexler, Eric, 233, 256
Dronamraju, Krishna, 16
Duesberg, Peter, 75

e-Biomed: The Journal of
    Regenerative Medicine, 163,
    164, 165, 229
E. coli, 68, 81, 89, 195, 246–47
Eckhart, Walter, 75

Edwards, Robert, 70, 71–72, 112, 127, 131–32, 149, 151, 152–53, 224

Eibert, Mark, 174

Eisenhower, Dwight D., 27–28

El Dorado Rejuvenation and Longevity Institute, 2, 48, 234, 235

Elixir Pharmaceuticals, 192–94

Elixxir, 232

Ellison, Larry, 231

Ellison Medical Foundation, 231

Engines of Creation (Drexler), 256

EPO, 247

ES Cell International, 148

Esfandiary, Feridouin M., 54–55. See also FM 2030

Essex, Max, 84, 85, 93, 95

Ettinger, Robert, 50, 52, 54

eugenics, 20–21

Evans, Martin, 114–15

Exeter Life Sciences, 235, 239, 242

expressionism, 16–17

Extreme Life Extension conference, 254–58

Extropian Declaration, 59

Extropian Principles, 59

Extropianism, 59–61, 195, 220–21, 251, 252

Extropy, 59

"Fabricated Babies" symposium, 127

Faloon, William, 232

Farley, Peter, 78

Fauci, Anthony, 161

feline leukemia virus, 84–85

First Immortal, The (Halperin), 232

Fleming, Donald, 70–71

FM 2030, 53–55, 57, 61; as a "neuro," 62; reason for name change, 55; and "21st Century Gatherings," 54

Foman, Robert, 82

Food and Drug Administration (FDA), 3

Foster, Roland, 146

Fountainhead, The (Rand), 59–60

Fox, Maurice, 31, 39, 68, 220

Francis, Donald, 77

Frankenstein (Shelley), 13, 124, 127

Franklin, Rosalind, 41

Frazer, Sir James George, 17–18, 136, 250

Fredrickson, Donald, 100

Freitas, Robert, 194, 198, 257

Friedmann, Theodore, 69, 96, 146, 213, 248

fruit flies, 25, 32–33, 104

Fukuyama, Francis, 128, 129–30, 147, 154, 195

Gage, Fred, 140

Galacidalacidesoxiribunucleic acid (Homage to Crick and Watson) (Dali), 131

Gallo, Robert, 77, 84–85, 93, 95

Galton, Francis, 20

Gearhart, John, 119, 120, 122, 148

"Gene," 142, 215

gene splicing. See DNA, recombinant DNA

"Gene Therapy for Human Disease?" (Friedmann and Roblin), 69

Genentech, 80–84, 89, 207, 220

"Generation in Search of a Future, A" (Wald), 69

genetic diseases, 69, 149, 152–53; and gene therapy, 97, 213

genetic engineering. See DNA, recombinant DNA

Genetic Savings and Clone, 241, 242

Genetic Systems, 83

Genetic Therapy, 97, 213

germline therapy, 152, 154

Gernsback, Hugo, 25

Geron, 103, 104, 111–12, 120–21, 123, 145, 181, 202, 214; etiology of name, 110; purchase of Roslin Biomed, 124

Gibson, William, 56, 57, 61

Gilbert, Walter, 44, 67, 68, 72, 73, 75, 79, 81, 94

Glaser, Donald, 78

Glenn, Paul, 231

Glenn Foundation, 231

Global Business Network, 58

Goedell, David, 80–81

Goffman, Ken. See R. U. Sirius

Golden Bough, The (Frazer), 17–18, 136, 250

Goldman, Bob, 2

Goldstein, Sam, 107, 187

Goodman, Ed, 83–84, 211–12

Greider, Carol, 108

Grove, Andrew, 145, 148

Guarente, Leonard, 192, 254

Gurdon, John, 71, 118, 175, 178, 180, 204

Haldane, John Burdon Sanderson, 11–12, 18–21, 25, 31, 45, 69, 148, 203, 248; as "the Communist, J.B.S. Haladane," 129; influences on, 15–17

Haldane's Daedalus Revisited (Dronamraju), 16

Halperin, James, 232

Hammond, John, 22–24, 71, 116, 124

Harley, Calvin, 103–4, 118. See also Geron

Harman, C. Mitchell, 240, 251

Harvard Psychedelic Research Project, 54

Haseltine, Florence, 29, 34, 35, 36, 43, 66, 68, 76, 90, 138, 211, 250–51

Haseltine, Jean, 29, 35, 37–38

Haseltine, William A., 9–10, 78, 123, 126, 159–60, 162, 163, 178, 181–82, 195, 241, 249, 254; and the ACT paper on cloning, 163–67; and AIDS research, 7, 84–86; childhood of, 33–38; college/graduate school career of, 42–45, 65–67, 69, 72–74; continuing belief in biotechnology, 218–21, 223–25, 253–54; at the Dana-Farber Cancer Center, 75–76; and Deeda Blair, 95–96; end of partnership with Venter's TIGR, 205–6; family of, 28–29; mantra of, 9, 209; meeting with J. Craig Venter, 6–9, 99–100; mother's influence on, 8, 37–38; 1972 paper for Nature, 74; postdoctorate research position, 74–75; and regenerative medicine, 181–83; reputation for bad behavior, 76–77; successful biotech career of, 206–12; and venture capital, 81–84; and Wallace Steinberg, 99–102. See also e-Biomed; Human Genome Sciences

Haseltine, William R., 28–29, 35–36

Hauser, Gaylord, 51, 55

Hawking, Stephen, 175

Hawthorne, Nathaniel, 147

Hayden, Charles Gervin, Jr. See Wicker, Randolfe ("Randy")

Healthcare Ventures, 88–90, 96, 97, 101, 206, 247

Heinlein, Robert, 48–50, 61, 72, 127, 178; "Future History" series of, 48; and Lazarus Long, 49–50

Hennenfent, Nick, 174

Heretics Society, 11

Hershey, Al, 39
Heward, Christopher, 226, 234, 239–40, 243
High Frontiers, 56, 57
Horowitz, Norman, 30
Houdini, Harry, 52
Howar, Nancy, 90, 94; and Wesley Associates, 94–97
Hubel, David, 66
Hughes, Karen, 143
"human-animal species," 139
Human Cloning Foundation, 174–75; board members of, 174
human genome: complexity of, 149, 248; and gene sequencing, 98; purported number of genes, 125, 208–10, 219
Human Genome Project, 4, 69, 98, 195, 207, 213
Human Genome Sciences (HGS), 98, 100, 101, 123, 125, 207–8, 210, 211, 212, 219, 220; board members of, 223
human growth hormone (HGH), 2, 47, 89, 234, 239, 247
Human Reproduction, 119
Hunt, Mark, 170
Hunt, Tracy, 170
Hurwitz, Jerard, 70, 73
Huxley, Aldous, 16, 19, 54, 124, 127–28
Huxley, Hugh, 40
Huxley, Julian, 16, 52
Huxley, Thomas, 14

Immortalist Manifesto, The (Elixxir), 232
in vitro fertilization (IVF), 71–72, 112, 114, 126–27, 251
Industrial Genome Sciences. *See* Diversa
Infigen, 142, 215
Insuraclone, 169
interferon, 94

Island of Dr. Moreau, The (Wells), 9, 14–15

Jacob, Francois, 67, 68
Jenkins, Nicole, 186
Johnson, Tom, 187, 190, 238
Jones, Howard, 127
Jones, Richard "Dick," 62
Journal of Experimental Medicine, 33
Juvenile Diabetes Research Foundation (JDRF), 142–43, 148

Kandel, Eric, 247
Kass, Leon, 126–27, 128, 129, 144, 147, 154, 157
Kass Commission, 147
Kekich, David, 231–32
Kelly, Kevin, 57
Kennedy, Alison. *See* Queen Mu
Kent, Saul, 3, 52–53, 54, 61, 62, 217, 232, 254. *See also* Cryo-Span; Life Extension Foundation
Kenyon, Cynthia, 189–92, 194–95, 238
Klatz, Ronald, 2
Kleid, Dennis, 75, 80–81
Kleiner-Perkins, 110, 111
Kleinsek, Don, 110
Kouchner, Bernard, 160
Krauthammer, Charles, 139, 147, 156, 216
Kristol, William, 128, 164
Kronos, 233–35, 242–43
Kronos Longevity Research Institute, 240
Kurzweil, Ray, 58, 253–54, 59, 60, 256–57; on MOSH, 58

la Mettrie, Julien Offray de, 13
Lacks, Henrietta, 108
Lang, Fritz, 18
Lanza, Robert, 165–66

Lasker, Albert, 92
Lasker, Mary, 92–93
Lasker Award, 92
Leary, Timothy, 54–55, 62; SMILE slogan of, 54
Lederberg, Joshua, 19, 67, 70, 229–30
Lemler, Jerry, 255–56
Lewis, Peter, 238
Life Extension Buyers Club, 232
Life Extension magazine, 232
Life Extension: A Practical Scientific Approach (Pearson and Shaw), 5
Life Extension Foundation (LEF), 3, 232; and the FDA Holocaust Museum, 3
Life Extension News, 183
life extensionist movement, 3–5, 9–10, 231, 252; concentration of in Texas and California, 109
Life, Liberty and the Defense of Dignity (Kass), 128
Lithgow, Gordon, 186–89, 194, 238
Long Now Foundation, 58
Longevity Boot Camp, 225–30
Look Younger, Live Longer (Hauser), 50–51
Looking Backward (Bellamy), 20–21
LSD, 54, 63
Luria, Salvador, 39, 40, 70

Machine Man (la Mettrie), 13
MacLeod, Colin, 32, 33
Mahoney, Florence, 48–49, 92, 126, 241
Man Into Superman (Ettinger), 54, 69, 154
Marchi, John, 173–74
Martin, Gail, 112
materialism, 13
Matter of Life, A (Edwards), 127, 132

Maxam, Alan, 67
Maximum Life Foundation, 231
McCarty, Maclyn, 32, 33
McHugh, Paul, 154–55
McKibben, Bill, 251
McKlintock, Barbara, 39
Megan and Morag, 121, 124
Medical Research Council (MRC), 22, 40
melanocyte stimulating hormone (MSH), 240
Melov, Simon, 188
Melton, Douglas, 141, 248–49; attempts to talk to politicians about stem cell research, 142–47
Memory Pharmaceuticals, 247
Mendel, Gregor, 21
Merkle, Ralph, 1, 2, 198
Meselson, Matthew, 67, 68, 72
mesmerism, 52
Methuselah's Children (Heinlein), 49
Metropolis (Lang), 18
"Mighty Mouse" project, 241
Minsky, Marvin, 58, 59, 60, 198
Missyplicity Project, 235, 241–42
Mizar, 59
molecular biology, 33, 39, 45–46, 70–71, 113–14, 115, 154
Molecular Biology of the Gene, The (Watson), 67
Monod, Jacques, 67
Mondo 2000, 57
Moravec, Hans, 58, 59, 149, 198
More, Frank, 131
More, Max, 58–60, 198, 226
Morgan, Thomas Hunt, 22, 24, 71, 109
MOSH (Mostly Original Substrate Humans), 58
Muller, Hermann, 25, 32, 40

nanotechnology, 1, 2, 59, 182, 233

National Institutes of Health (NIH), 7, 27, 215; National Institute on Aging, 92, 107
National Review, 156
National Science Foundation (NSF), 27, 244
Nature, 14, 41, 74, 121, 168
Nature Biotechnology, 165, 206, 235
Neuromancer (Gibson), 56–57
Niehans, Paul, 48–49
Nosferatu, 17

O'Connor, Max. *See* More, Max
"On Living the Biological Revolution" (Fleming), 70–71
Optimism One (Esfandiary), 55
O'Reilly, Hunter, 131
Origin of Species (Darwin), 20, 21
Our Post Human Future (Fukuyama), 129; Haseltine's review of, 224
Oxford Biosciences, 110

Parkinson's Foundation, 148
parthenogenesis, 24–25
Pauling, Linus, 31, 40, 55, 109, 138
Pearson, Durk, 5–6, 10, 55–56, 183, 195, 227–28
Pedersen, Roger, 119, 120, 141
Pence, Gregory, 174
"Penultimate Trump, The" (Ettinger), 50
Pharming, 215
physics, 30–31
Piel, Gerard, 42
Pimentel, George, 43, 44
Pitts, Joe, 216, 217
planaria, 24, 71
Plant Genome Sciences, 101, 206
Poste, George, 99, 101
PPL Therapeutics, 121, 181, 215, 242
preimplantation genetic diagnosis (PGD), 150–52, 153

Primo 3M+, 198
Proceedings of the National Academy of Sciences, 79
Prolinia, 242
Prospect of Immortality, The (Ettinger), 52
Ptashne, Mark, 67, 72, 81

Quarles, Miller, 109–10, 258. *See also* Cure Old Age Disease Society
Queen Mu, 56, 57

Rael, 167–71
Raelians, 168, 170; and Dolly as a sign of the Second Coming, 168, 169, 170
Ramsey, Paul, 127
Rand, Ayn, 59–60
Redesigning Humans (Stock), 152
regenerative medicine, 181–83
"Regenesis" project, 246
Reich, Steve, 131
rejuvenation clinics, 12, 48–49
Reliance Life Science, 148
Repifermin, 182–83, 211
Reproductive Biomedicine Online, 151
Reproductivecloning.net, 175
retroviruses, 77, 84–85
Rich, Alexander, 44–45
RNA, 44, 67, 79–80, 104. *See also* DNA
"RNA Tie Club," 44
Robertson, John, 174
Roblin, Richard, 69, 96
Rockefeller, John D., 22
Rockefeller Institute for Medical Research. *See* Rockefeller University
Rockefeller University, 22
"Roger Moorgate." *See* Byrne, James
Rose, Michael, 187, 198–99, 254
Rosen, Craig, 100, 209, 220
Roslin Institute, 117–18, 121

Rossetto, Louis, 57
Rove, Carl, 143
Rubinstein, Bradley, 131
R.U.R. (Capek and Capek), 18, 81
R. U. Sirius, 56–57, 61, 63, 157
Ruvkun, Gary, 238

Safire, William, 140–41
Salk, Jonas, 42, 85
Sanger, Frederick, 67, 79
Sassoon, Sigfried, 15–16
Scheid, Al, 82–83, 210
Schrodinger, Erwin, 30–32, 102, 108, 203
science: in the Age of Reason, 12; and the Industrial Revolution, 13, 21–22; institutionalization of, 45–46; mainstream science, 3; post–World War II conservatism of, 63; and "scientists," 13; and World War I, 15–18
"Science: The Endless Frontier" (U.S. Office of Scientific Research and Development), 27
Scientific American magazine, 25, 42, 163, 166
Second Annual Conference on Anti-Aging Medicine, 1–6; and the "Infinity Award," 4
Second Annual Conference on Regenerative Medicine, 159
Second Creation, The (Wilmut, Campbell, and Tudge), 161–62
Segre, Emilio, 44
Selfish Gene, The (Dawkins), 60–61
Shaw, Sandy, 5–6, 55–56, 183, 195, 227–28
Shay, Jerry, 108
Shelley, Mary, 13, 137
Short, Jay, 90
Simpson, Joe Leigh, 150–51, 151–53
Singapore Biomedical Center, 148
Sleeper (Woody Allen), 53

"smart drugs," 56
Smith, Christopher, 163–64
Smith, F. E., 55
Smith-Kline Beecham, 99, 101, 235
Smithies, Oliver, 115
Soros, George, 238
Spemann, Hans, 118
Sperling, John, 176, 230, 233, 235–43; funding of cloning projects, 235. *See also* Kronos
spiritualism, 52
Spurway, Helen, 25
Startling Stories, 50
Steinberg, James, 161
Steinberg, Wallace, 86–90, 111, 126, 192, 203, 205, 210, 212, 213, 220; and Deeda Blair, 94–97; and William A. Haseltine, 99–102. *See also* Healthcare Ventures
stem cells, 114–16, 117–18, 120, 128, 122, 137, 201–5; development into cardiomyocetes, 201; lack of idea how to make them useful to people, 213–15; lines available for scientific research, 144; Tony Blair's endorsement of ES cell science, 141
Steptoe, Patrick, 71, 112, 127, 131–32
Sterns, Cliff, 139–40
Stice, Steve, 242
Stock, Gregory, 152, 240
"Structure for Deoxyribose Nucleic Acid, A" (Watson and Crick), 41
Stump, David, 220
Sulston, John, 185
Swanson, Robert, 80

Tatum, Edward, 29–30, 70
Technology Entertainment Design (TED) conference, 58
Telespheres (Esfandiary), 55

telomerase, 104, 108
teratomas, 112–13
The Bioethics Project, 128
The Foresight Institute, 58
The Institute for the Future, 58, 243
The Institute for Genomics Research (TIGR), 100, 101, 205–6
The Singularity, 61, 149, 256–57
Theosophy, 52
Thompson, Tommy, 142, 217
Thomson, James, 120, 122–23, 202, 204
Time Enough for Love (Heinlein), 50, 72, 178
tissue engineering, 214–15
T. O. Morrow, 59
transdifferentiation, 179, 180–81
transhumanism, 14, 51–52, 60, 61, 153, 154, 156, 244, 252; argument for life extension, 194; and human add-ons, 181
TransVio Technology Ventures, 231
Travers, Andrew, 73–74
Turney, Howard, 47–48, 235; as Lazarus Long, 48
21st Century Medicine, 232, 233

Ullis, Karlis, 226
Upwingers (Esfandiary), 55
Upwingers, 55
Urey, Harold, 44
US News and World Report, 163, 166
U.S. Office of Scientific Research and Development, 27

Valiant Venture Ltd., 169
Venter, J. Craig, 7–9, 98–100, 126, 190, 205–6, 208, 213, 220, 248. See also Celera
venture capital, 78, 81, 89
Verma, Inder, 68, 96, 104, 213
Verne, Jules, 13–14

Viagen, 242
Vinge, Verner, 61
virtual reality (VR), 58
Vita-More, Natasha, 55, 60, 198, 226
Vorilhon, Claude. See Rael

Wakayama, Teruhiko, 147, 216, 217
Wald, George, 69
Walford, Roy, 4, 107, 198
Wall Street Journal, 82, 166
Walton, Alan, 110
Watson, James, 7, 8, 40–41, 44, 67, 70, 72, 73, 75, 76, 149, 212; on human cloning, 72, 135, 224; and the Human Genome Project, 98; paper on DNA, 41
Weissman, Irving, 140, 163
Weissmann, Charles, 73
Wells, H. G., 14–15, 61, 203
Werner, Hal, 88, 96, 98, 219
West, Michael, 87, 103, 106, 108–12, 118, 120–21, 123–24, 126, 162, 163–67, 183, 195, 198, 215–18, 229, 254, 258; childhood of, 104–5; Ph.D. work, 107–8; study of religion, 105–6. See also Advanced Cell Technology; Geron
What Is Life? (Schrodinger), 30–32
Whewell, William, 13
White, Ray, 68
Whole Earth Review, 57
Wicker, Randolfe ("Randy"), 171–78. See also Clone Rights United Front
Wilkins, Maurice, 41
Will, George, 140
Williams, Ted, 62
Wilmut, Ian, 116–17, 121, 134, 140, 161–62, 163, 242
Wilson, E. O., 189, 249
Wired, 57, 168

Wistar, Isaac Jones, 22
Wistar Institute of Anatomy and
    Biology, 22
WISTRAT, 22
World in 2030, The (F. E. Smith),
    55

Wright, Woodring, 108

Yarborough, Don, 109
yogurt, as antiaging health food, 51

Zavos, Panos, 145–46, 167–68